Benelli Road Racers

RAYMOND AINSCOE
with
GIANNI PERRONE

Published in 1995 by Ilkley Racing Books
3 Mendip House Gardens Curly Hill Ilkley LS29 0DD

Copyright © Raymond Ainscoe and Gianni Perrone 1995

All rights reserved. Apart from any fair dealing for the purpose of private study, research, criticism or review, as permitted under the Copyright, Designs and Patents Act, 1988, no part of this publication may be reproduced, stored in a retrieval system or transmitted in any form or by any means, electronic, photocopying, recording or otherwise, without the prior written permission of the publisher. All enquiries should be addressed to the publisher.

Benelli Road Racers: ISBN 0 9524802 0 4

Printed by Amadeus Press Ltd., Huddersfield.
Typesetting by Highlight Type Bureau Ltd., Shipley.

Front cover: Renzo Pasolini, 500 cc Benelli at Quarter Bridge, 1967 Senior TT (Nick Nicholls).

About the authors:- Raymond Ainscoe contributes articles on Italian racing motorcycles to "Classic Racer" and "Classic Legends" magazines. His previous books are "Gilera Road Racers" and "Laverda". An enthusiast of the Italian classic bike scene, having competed twice in the revived Giro d'Italia, he enjoys participating in street circuit events such as the Cattolica Motor Meeting and Treviso's Circuito delle Mura. Other enthusiasms are his three children, the football of Manchester United and the music of Hector Berlioz.

Gianni Perrone, a resident of Rome, was a professional racer throughout the 1960s and 1970s, riding a wide range of machinery from a 125 cc Rumi Junior to a 750 cc Yamaha. He has two victories in the Greek Grand Prix, an 8th place in the Italian Grand Prix at Monza in 1970 and numerous club championships to his credit. Today he participates in classic events throughout Europe and has added journalism to his skills, writing for the Italian magazines "In Moto" and "Legend Bike". Gianni's wife and three sons do not believe that he has any interests other than motorcycles.

CONTENTS

Acknowledgements

Foreword by Dorino Serafini

1. The Pioneer Years
2. The Pre-War Years
3. The Ambrosini Years: 1945-1952
4. The 250 cc Single Reborn.
5. Post-War Miscellany: The Leoncino, Motobi, 50 cc Racers.
6. The 250 cc Four: 1962 - 1969
7. 350 cc Exploits: 1965 - 1973
8. The 500 cc Campaign: 1966 - 1973
9. Postscript.

ACKNOWLEDGEMENTS

In preparing this book, I have been extremely fortunate to have been able to meet, or correspond with, members of the Benelli family and the majority of the surviving riders, engineers and mechanics who played parts, large or small, in the saga and to all of them I offer my sincere thanks for their assistance.

There are, I fear, too many such contributors to name each and every one but I must record my particular appreciation of the enthusiastic help of the following. First and foremost, for scouring his memory of the pre-War years and for kindly writing the Foreword, Dorino Serafini, European champion of 1939 and still a resident of Pesaro.

For their tales of Ambrosini's endeavours, his colleagues Bruno Zoffoli and Natalino Bersani, and Africo Paolucci, all from the great champion's home-town of Cesena. And for his recollections of the Motobi and Benelli race-shops of the 1950s and 1960s, former racer and technician Primo Zanzani, now plying his expert trade as a brake manufacturer just outside Pesaro.

Thanks also to Umberto Masetti, one of life's great story tellers; Giancarlo Morbidelli, an inveterate Benelli fan, and his cousin Fernando Bruscoli of the Benelli Club of Pesaro.

Many of the photographs have been provided by those stalwarts of their trade, Nick Nicholls, Mick Woollett, Michael Dregni, Sandro Colombo and my friend Sauro Rossi of Boretto, who boasts one of the most fascinating archives in Italy. To them all, my thanks. My appreciation goes to author, Italophile and friend Tim Parker who encouraged me to write the story.

A particular debt of gratitude is owed to Gianni Perrone, formerly a Continental Circus performer of repute and now a leading authority, participant and journalist on the classic scene. He is the very soul, the embodiment, of an enthusiast. Without his unstinted hospitality and generosity, not to mention the provision of photographs from his marvellous collection, this book could not have appeared in this form.

Finally, thanks to my wife Elizabeth who has accepted my jaunts to Pesaro with endless patience.

Raymond Ainscoe
Ilkley, January 1995.

The Benelli brothers: from left to right, Tonino, Francesco, Giovanni, Giuseppe, Filippo and Mimo *(Benelli)*.

FOREWORD by Dorino Serafini

I began with Benelli in 1927 by which time the factory's products were two models: the 125 cc and 147 cc two-strokes.

The 175 cc four-stroke soon appeared, based on an original design by Engineer Giuseppe Benelli who built an engine with an ohc assembly driven by a gear train.

The bike was an instant success and finished on the leaderboard not only in national events but also in the international field. It was fast, handled well and had excellent road-holding.

I started with Benelli as a test rider. In those times, there were no test benches; the engines were tried on the open roads.

Subsequently I became part of the race team together with Tonino Benelli and Carlo Baschieri. Tonino was a lively, intelligent man; a good example for the youngsters. He knew how to get the best out of his 175 cc bike, adapting his speed to the track and the position of his rivals in a race. He was fearless and at the same time prudent; he was not afraid to take a risk but only if it were necessary. He was the complete master of his machine.

He took part in numerous races, winning or being placed in many of them. He taught me a considerable amount which enabled me to win races. I remember one particular episode. I won a championship race at Lugo di Romagna when Tonino was forced to retire; but he was absolutely delighted that I had brought a Benelli to victory. These years were merely the start of a marvellous history for Benelli racers.

Dorino Serafini
Pesaro

The youthful Tonino adopts a racing pose.
(Benelli)

Chapter One: The Pioneer Years

Arguably the most striking feature of the growth of the Benelli factory from its humble beginnings in 1911 was that its guiding light in its formative years was no high-handed aristocratic despot in the Count Agusta mould nor a rider-cum-mechanic such as Giuseppe Gilera. No, instead it was an impecunious widow, Teresa Benelli, whose initiative and enterprise drove the business on from the most unprepossessing of origins.

Signora Benelli was widowed in 1907 and thus suddenly found herself responsible for the well-being of an extensive dependent family, including her daughters, Pia and Maria, and her six sons, Giuseppe, Giovanni, Francesco, Filippo, Domenico and Antonio, all of whom were to play a part in the story that was to unfold in the years to come.

The resourceful widow nurtured ambitious plans for her brood. Both Giuseppe and Giovanni were despatched to the Industrial Institute in Fermo, some 70 miles away, to study engineering. Thereafter Giuseppe obtained a university degree while Giovanni served his apprenticeship with the Fiat and Bianchi concerns.

However, they were summoned back to Pesaro, for in 1911 their mother ventured into a business which operated from the family home in via Mosca in the heart of the historic town. Initially, the fledgling company was called "l'Officina Meccanica di Precisione f.lli Benelli". Signora Benelli ran a tool workshop, directing half a dozen outside employees and five of her six sons. The youngest - known as Tonino - was still at school, having been born in 1902.

Under the supervision of the enterprising matriarch, the business flourished until an unforeseen disaster struck in 1916 when an earthquake devastated the centre of the Adriatic town, destroying the family's premises and many of their machines. At this time both Giuseppe and Giovanni were working in Milan and, reluctantly, Signora Benelli decided to uproot her family and transplant them to the Lombard capital.

At the last moment, a much needed slice of luck fell her way when a relative offered assistance in acquiring suitable premises in via Mameli on the edge of Pesaro. The site had formerly been occupied by the Molaroni brothers who had produced a series of two-stroke single cylinder engines of their own design but they had recently moved out. Hence, the Benelli business - soon to be renamed "la Fabbrica Motocicli f.lli Benelli" - moved to the property that it was to occupy for another 65 years. By contrast, history records that Moto Molaroni faded into oblivion in the mid-1920s.

Although the precise date has been lost, just after the Great War, probably in 1919, Giuseppe and Giovanni produced their first motorcycle engine. It was an unassuming 75 cc two-stroke affair that was intended to be clipped onto the front forks of a bicycle in time-honoured fashion. Such primitive practices had long-since been forsaken by the English manufacturers but their Italian counterparts lagged behind in a number of design techniques by a good 10 or 15 years.

Needless to relate, the brothers soon discovered that a flimsy bicycle frame was not of sufficient substance to endure the punishment meted out by their tiny power plant. Accordingly, in 1920 they went the whole hog and designed their first motorcycle, featuring two speeds and a chain final drive. Much play was made of this latter feature which, in Italy, was considered to be relatively avant-garde. The little factory produced a range of two-stroke engines, of 75 cc, 98 cc, 125 cc and 147 cc, to offer to its public.

During the pioneering years, Giovanni was an enthusiastic competitor in the burgeoning sport of the times. He participated in the local hill-climbs and the cross-country runs that were such a prominent feature of the contemporary sport, aboard a souped-up version of the 147 cc machine. In common with other marques at that time, Benelli did not build genuine racing motorcycles; instead Giovanni raced what was essentially a standard production machine that merely enjoyed the benefit of one or two fairly predictable tweaks such as low handlebars and an extended saddle.

It was however to be the youngest brother, Tonino, whose competitive exploits were to thrust the marque to the fore. Aboard the humble 147 cc steed Tonino made his debut at Pesaro in 1923 when he was a praiseworthy second behind established star Primo Moretti on a 500 cc Moto Guzzi. Before season's end, he had entered a race at Monza, the newly-opened purpose-built circuit in the royal parkland north of Milan, and he took a fifth spot.

Tonino's first victory came in the following year, at Parma, and in the process he beat Giovanni who was thus hastened into premature retirement, preferring henceforth to fettle his brother's machines. Tonino notched another success at Bergamo before the year was out.

It was perhaps just as well that Tonino's energies were to be channelled into racing. One of his early escapades arose from his acceptance of a challenge proferred by a local mechanic, the proud owner of a 350 cc AJS, to a race around the streets of Pesaro, recorded as taking place on 11th March 1925. Tonino duly won the wager but as he crossed the agreed finishing line he was apprehended by the constabulary who had been forewarned of the prank. Inevitably, a hefty fine was levied to penalise the flagrant breach of the town's speed limits. It should be mentioned that the police were never tardy in imposing such fines, as they were entitled to a percentage to supplement their wages!

Back in law-abiding competition, Tonino's victories at Pesaro and Senigallia in 1925 and 1926 gave but scant hint of the shape of things to come.

Tonino's spectacular breakthrough came in 1927 when a 175 cc four-stroke machine was designed by Giuseppe. With a salesman's eyes to advertising the family's wares, Giovanni developed one of these bikes for his brother's use. The new racer achieved immediate

THE PIONEER YEARS

success on May Day at the Circuito dei Tre Monti at Imola, leading the 100 mile race that counted towards the national 175 cc title from start to finish.

At the end of July, Tonino entered and dominated his class in one of the two domestic classics, the Italian TT over the 24 mile Lario circuit. With its three hundred bends, demanding descents and ascents, it was an unashamed copy of and answer to the Manx Mountain course, and was held in equally scenic countryside between the southern legs of Lake Como, otherwise known locally as Lake Lario. After Tonino's success it was written of him: "He is the master of his machine and rides, one might say, with his head. He is not a wrecker; he is a serious rider who risks only the possible and knows exactly what he can do. We studied him on the descent from Ghisallo and we can confirm that he has one of the cleanest cornering styles. Perhaps his excellent machine allows him to descend with more confidence than many others but certainly Tonino must be considered a class rider."

By the end of the season Tonino had captured the other classic, the GP at Monza, and seven of the eight title rounds, thereby appropriating the national 175 cc title and earning himself the right to wear the prestigious tricolour jumper accorded to each title-holder for the forthcoming season. He rounded off the year with a record-breaking spree at Monza, taking world records for 50 km., 100 km., 100 miles and one hour, the last at about 78 mph.

The authoritative daily sports journal, La Gazzetta dello Sport, that had been instrumental in promoting the sport in Italy in the years before the Great War, concluded its review of the season with the comment that "of the champions of Italy, two seem to be at the pinnacle: Tonino Benelli and Luigi Arcangeli. The first has strung together an extraordinary run of victories and with his tireless energy has established a significant advantage over all his rivals."

For 1928 Tonino's virtually unassailable bike was finished in traditional Italian red. It was powered by a 172 cc engine, of 62 x 57 mm dimensions, which featured an external flywheel and twin exhaust pipes. The two valves were controlled by a sohc assembly, with exposed coil springs which were subsequently replaced by hairpin springs. An L-shaped gear train was employed, running on the right hand side of the engine, with a gear driving the magneto sitting at the front of the crankcase.

Other components were a separate three-speed Albion gearbox, an oil tank beneath the saddle and conventional cycle parts: a diamond-pattern tubular frame, girder front and rigid rear forks. Dry weight was less than 200 lb. and the engine's 10 bhp at 6,500 rpm permitted a top speed of over 80 mph.

1928 began brightly, for Tonino's victories at Messina and Turin were merely the prelude to a repeat success over the daunting Lario course. Once again a reporter lavished fulsome praise on him for his display in the Italian TT: "I saw Tonino go through the curves on the

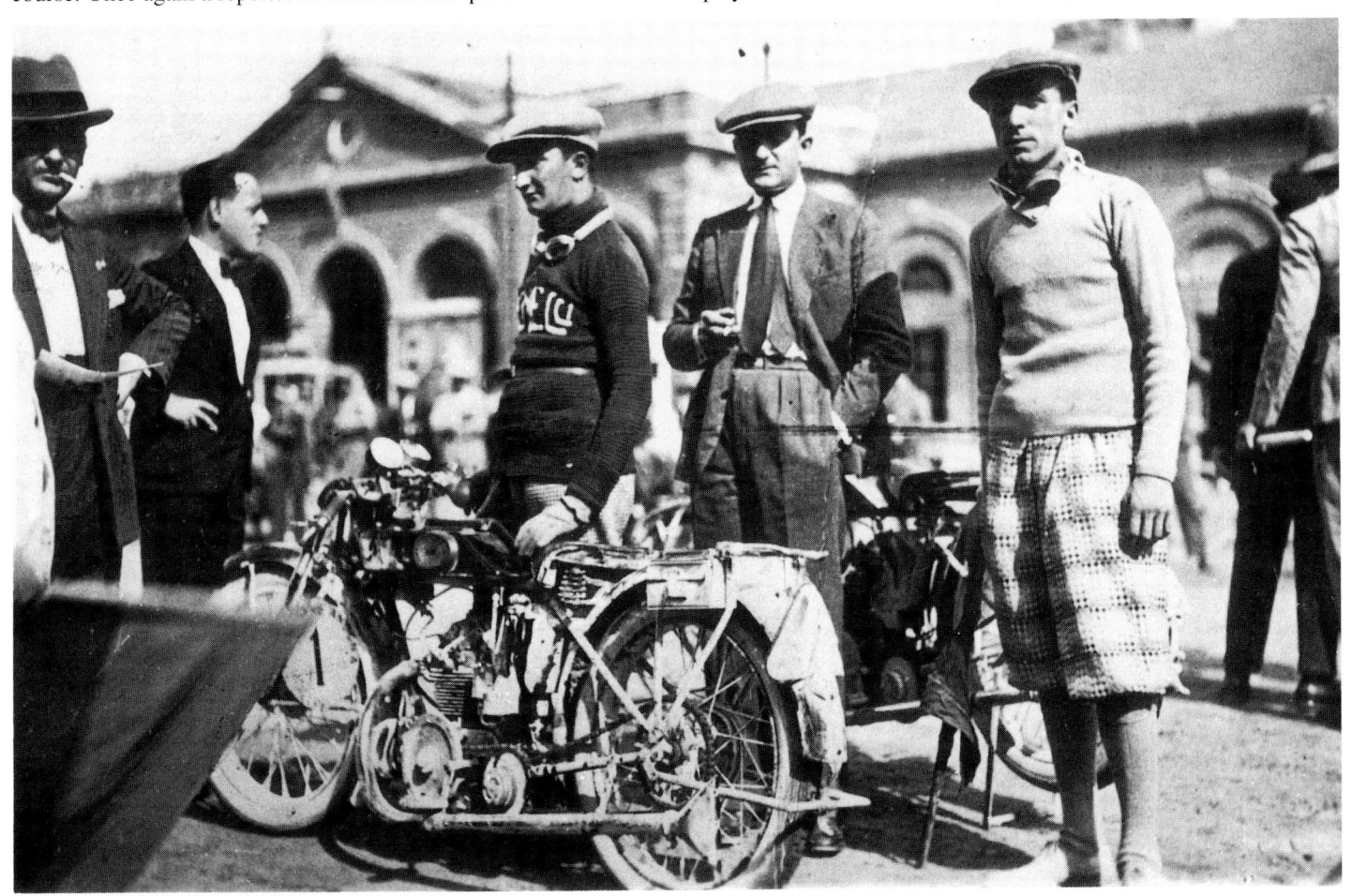

Tonino pictured at the Turin Grand Prix in 1928 with his 175 cc model *(Benelli)*.

descent from Barni and frankly his riding style was perfection; it seemed that he followed an ideal line. The Benelli is a phenomenal motorcycle but young Tonino is a phenomenal rider who can now aspire to perform abroad successfully."

On the continent the 175 cc class was an established racing category and featured in the European championship that had been initiated in 1924 and was determined on the results of a solitary Grand Prix. In 1928, the honour of hosting the event fell to the Swiss GP, held that year in Geneva.

Encouraged by their domestic successes, the Benelli factory entered the race but unfortunately Tonino suffered a nasty tumble. Team-mate Brusi, another native of Pesaro, salvaged second spot behind countryman Alfredo Panella aboard the rapid side valve Ladetto e Blatto from Turin, a barely remembered marque that has subsequently all but disappeared into the mists of time.

Although Tonino once again secured the national title, the GP at Monza was a disappointment. Having run short of fuel, he trailed in a distant fourth, behind Artur Geiss aboard his supercharged DKW, the other dominant model in the class at the time, and a brace of Ladetto e Blatto racers.

Tonino raced rarely in 1929 but the year was far from barren for the factory, thanks largely to the efforts of second string Riccardo Brusi. Brusi's career began in 1925 when he competed in the Giro d'Italia aboard a 350 cc AJS. He joined Benelli in 1927. His golden opportunity came in 1929 when he was promoted to team leadership, and he certainly grasped it with both hands, notching eight victories, including the Royal GP at Rome. He also managed the runner-up position in that year's European championship by dint of his second spot in the Spanish GP in Barcelona, behind one of the potent DKWs ridden by Klein.

The year ended momentously for the factory at the GP at Monza. Brusi, by now the established favourite, confirmed his status by shooting into a seemingly unassailable lead only to roll to a halt on lap 6 out of 20. Geiss inherited the lead when Tonino retired on lap 17 but came under substantial pressure from Carlo Baschieri aboard the third Benelli. Baschieri just caught the Deek on the line to take the chequered flag by the proverbial whisker.

Sadly for Benelli, Brusi's progress had not gone unnoticed and he was signed up by the rival Moto Guzzi concern. As a counterweight, Tonino returned with a vengeance in 1930, collecting the Coppa del Mare at Leghorn, the Targa Florio at Palermo, the Royal GP, the Lario TT and his third national title.

By 1930 racing was becoming more professional in outlook; the Continental Circus was well established, manufacturers were building genuine racing motorcycles and substantial teams were entered in the major events. For instance, Benelli entered four machines in the GP at Monza; Tonino, restored to peak form, won yet again, Miele was third while Raffaele Alberti and Baschieri retired. Another example of the increasing organisation surrounding the sport was that Domenico Benelli - known as Mimo - was now assuming the role of team manager that he was to occupy for almost four decades.

Meanwhile, Giovanni was not content to rest on his laurels; he was anxious to promote a new production 175 cc model and set about updating the racer. The engine's dimensions became square at 60.5 mm, increasing the capacity to 173.8 cc, while a dohc assembly was introduced, so that the gear train was T-shaped. A puny radiator was added to the front of the oil tank and the tyres became fatter at 2.75 x 21 in., while weight crept up by about 6 lb. It was to be this machine which was to thrust Benelli to the forefront of international competition over the next few years with GP victories in Switzerland, France, Belgium and Holland.

Tonino riding the Lario circuit, 1928
(Sandro Colombo).

THE PIONEER YEARS

In 1931 Pesaro's newly formed Moto Club, with Tonino designated as the "Technical Director" and Brusi the grandiosely named "Director of Sport", organised a race over a circuit along the seashore and around its port area.

Benelli's honour was upheld in the race by a new recruit to the 175 cc class, Dorino Serafini. Serafini was born in Pesaro in 1909 and his father, a Douglas owner, knew the Benelli brothers. He talked them into offering Dorino his first job as a tester in the factory. Serafini recalled those days: "I rode all the factory's production machines for fourteen hours a day, starting at seven in the morning. We used the open roads around Pesaro, both inland and along the coast. I also tested the racers and we would be given three frames and three engines. We would juggle with them until I could find the best combination that Tonino could then use in a race. My speciality became carburation; in fact, every year Mr. Binks would visit the factory and, on more than one occasion, he only half-jokingly offered me a job in England."

In the main, 1931 represented another page in the saga of Tonino's dominance aboard his dohc model. Once again, a fall at the Swiss GP, this time at Berne, interrrupted his triumphal progress, allowing Baschieri to take the flag. That mishap apart, it was a season in which Tonino demonstrated his mastery, with victories at Forli, Lario, Rome and Leghorn contributing to his fourth national title.

The Benelli squad virtually monopolised the leaderboard of the Italian GP at Monza for Tonino was backed up by Alberti, Miele and Baschieri in third, fourth and fifth places respectively.

1932 started auspiciously at the Italian GP, designated as that year's European title race, held over the Littorio circuit to the north of Rome, having been moved from Monza reputedly for political reasons, prompted by Mussolini. Il Duce was undoubtedly a great fan of motorcycle sport, holding membership card number one for the Italian Motorcycle Club, and he attended many events. Indeed, this race, around the Littorio airfield, was honoured by the presence of the King.

Benelli made a tremendous effort, entering four bikes, sporting a crude fairing over the front forks and with the rests moved back to the rear axle. Serafini led in the early stages but soon retired, as did new boy Borghese. Baschieri won, despite a pit spot to clear a lubrication problem, while Tonino was for once outridden and relegated to second place. Indeed, it was very much Baschieri's season, for he also beat Tonino to the national title.

Serafini was also coming to the fore with victories at Macerata, Lugo, Ancona and Forli to his credit. At the end of the season, he was poached by the rival MM squad, although years later Serafini admitted that his move to the Bologna factory had been motivated not by professional ambition but by an affair of the heart which was in danger of becoming a local scandal!

In October 1932, Tonino celebrated the last of his 34 victories, leading the field home by over two minutes at the French GP at Montlhery. However, his competitive career ended on 27th November when he crashed in torrential rain at the Tigullio circuit, not far from Genoa, breaking a leg, an arm and ribs and puncturing a lung. His brothers were devastated when they arrived at his bedside and for fifteen days he lay unconscious while the doctors despaired for his life. When he regained consciousness, legend has it that his first words were to enquire who had won the race. Inevitably, it had been a Benelli, ridden this time by yet another youngster, the Milanese

The works sohc 175 cc racer of 1929/30 vintage *(Benelli)*.

Carlo Baschieri, Tonino and Giovanni Benelli at the Rome Grand Prix, 1930 *(Benelli)*.

THE PIONEER YEARS

jockey Giordano Aldrighetti, who was to move on to successes with an Enzo Ferrari-sponsored Rudge and then Gilera before meeting his death at the wheel of a Ferrari in 1939.

When Tonino returned on the train to Pesaro station he found that thousands of his townsfolk had turned up to greet 'The Flying Benelli'. After a slow recovery, it was obvious that his racing days were over, and so he took his place alongside his brothers in the factory.

Shorn of Tonino's unique skills, 1933 was a relatively bleak year for the team on the tracks, and old boy Serafini hijacked the national title with his MM. Francesco Lama, who had made the reverse journey, had the consolation of a victory at Alessandria on his tiny Benelli and the team's number one, Alberti, appropriated three victories of the highest quality; the long-distance Milano-Napoli marathon, the Dutch TT and the Belgian GP.

Stung by the reverse on the national front, Giovanni produced a revised version for 1934 with a four speed gearbox and pedal change, a larger oil tank and 15 bhp at 9,000 rpm which gave a top speed of 90 mph. The recruitment of a top class rider in Amilcare Rossetti did the trick and he regained the national title with seven victories including the Milano-Napoli and Lario events.

Second-string Raffaele Alberti upheld the marque's repute with victory in the prestigious Coppa Acerbo. This classic event was held over a rapid triangular circuit based on the old fortress town of Pescara and it was named by one of Benito Mussolini's Cabinet Ministers, Professor Giacomo Acerbo, in memory of his brother Captain Tito Acerbo who had been killed in the Great War.

Possibly the marque's most notable successes achieved during that season were those in the Low Countries by the Dutch privateer Ivan Goor. Goor was already handsomely into the veteran stage, nudging fifty, when he won the 1930 175 cc European championship by winning the Belgian GP at Spa, on his DKW.

Goor acquired a Benelli for the 1934 season, although it was to be maintained on a private basis, and he took a string of wins at Chimay, Floreffe, Spa, the French GP at Montlhery and the Dutch TT at Assen - the last of which was that season's European title decider. Eric Fernihough's magneto trouble enabled Goor to take the flag ahead of the Belgian rider Dickwell - yet another who would feature again in the Benelli history - aboard his own creation, a Barbe.

At that stage, with the national and European titles in the safekeeping of Benelli, the 175 cc category was stripped of its championship status. But the brothers were determined not to rest on their hard-earned laurels; they decided to transplant their wealth of expertise to the 250 cc category and a glorious new chapter was about to begin.

A single camshaft 175 cc racer forming part of the collection of Giancarlo Morbidelli.

Amilcare Rossetti, Lario 1934. *(Sandro Colombo)*

Chapter Two: The Pre-War Years

With the sharpest of axes falling precipitously on the title status of the 175 cc category, as of necessity Giovanni turned his attention to a 250 cc racer that was a logical development.

The Pesaro factory's first 250 cc bike was, in truth, merely an enlarged version of the trusty 175 cc machine. The engine's dimensions were extended to 67 x 70 mm for 246.7 cc and the cycle parts were more robust than those of its predecessor. The twin exhaust pipes were retained, initially. Weight was, probably over-optimistically, reported at 230 lb. (Guzzi no doubt treated this claim with a hefty pinch of salt, in the knowledge that their own racer tipped the scales at 270 lb.) The motor's 25 bhp at 8,000 rpm propelled the machine to a top speed, in race trim, of about 105 mph.

The new machine was race prepared for the 1935 season and lightweight jockey Raffaele Alberti was selected as the team's number one rider for the occasional appearances that were planned for the year ahead. His first task, however, was to tilt at the world flying kilometre record. The scene of the attempt was the Firenze-Mare autostrada at Altopascio. With minimal refinements, consisting merely of discs over the rear wheel spokes and a small petrol tank, Alberti established a mark at over 113 mph that also appropriated the record for the 350 cc class into the bargain.

The new racer, together with its 500 cc sister, was then shipped to the Italian Protectorate of Libya to compete in the Tripoli GP that was one of the social events of the year and attracted all the leading Italian teams. Against the full might of the Guzzi squad, Alberti raced honourably but was constrained to retire.

At the end of May came the Italian GP that had suffered a unique downgrading in status. The sorry tale was that Il Duce's territorial designs on hapless Ethiopia had rendered diplomatic relations extremely fragile with the consequence that the foreign teams opted not to race at the meeting. Thus relegated to national status, the event was re-named, in grandiose Italian fashion, the Monza Grand Prix of the Royal Italian Motorcycle Club.

The domestic manufacturers supported the event enthusiastically. Benelli recruited their old rival, the ex-Ladetto e Blatto teamster Panella, but he soon retired the second string bike. Alberti lasted until lap 15 out of 25, having duelled with the ultimate victor, Aldo Pigorini aboard his Guzzi.

1936 was very much a case of more of the same for the Benelli team: participation in occasional events without conspicuous success. The Milanese star Giordano Aldrighetti, who had ridden Enzo Ferrari's Rudges in the interim, returned to the fold on a part time basis and scooped the GP of Geneva at the end of June - a sign that the Benelli was at least competitive.

The highlight of the year on the international front was the European GP allocated to the German event at the Sachsenring. At half-distance Guzzi's "Black Devil" Omobono Tenni wrecked his clutch and gearbox so that Alberti inherited the lead. Sadly for the Pesaro team, on lap 28 out of 35 his Benelli cried enough. Hence, Tyrell Smith moved into a lead that he was never to relinquish and, unexpectedly, his unsung Excelsior took the European title.

In September the teams converged on Monza for the Italian GP - once again an international. The riders indulged in a version of musical chairs, with Aldrighetti opting to ride in a fearsome Guzzi trio alongside Tenni and Amilcare Moretti. To replace him, Benelli recruited Amilcare Rossetti and a youthful Nello Pagani who had already made his mark with the Italian 250 cc title in 1934 on a humble Miller.

For this, the most prestigious of Italian events, Giovanni Benelli introduced a couple of modest refinements to his bikes. The twin exhaust pipes were scrapped in favour of one extremely long pipe on the left hand side, and rear springing, of the plunger variety, was belatedly introduced. Alas, the modifications were of no avail. Aldrighetti, Tenni and Moretti romped home in first, second and fifth places respectively, with Pagani, Alberti and Rossetti managing third, fourth and sixth.

This unacceptable humiliation signalled the end of the road for the first generation lightweight Benelli and Giovanni was prompted into a complete re-design. Accordingly, there were no factory entries in 1937, although a former works bike gave newcomer Enrico Lorenzetti his first taste of success at Mendrisio, just over the border in Switzerland.

All in all, 1937 was a bleak year for the Benelli family, with the tragic loss of its talisman, Tonino. The youngest of the brothers, although destined never to race again after his devastating Tigullio crash, could not resist test riding the factory's new models. It was on one such run, returning from Pesaro from a trip up the coast to Rimini, that, in evading a lorry, he crashed fatally near Riccione on 27th September. His funeral was attended by thousands of his mourning townsfolk.

The re-drawn 250 cc machine, devised by Giuseppe and Giovanni in tandem during a lengthy gestation period, was ready for the start of the 1938 campaign. Although it bore a distinct resemblance to its predecessor, it was in fact virtually an entirely new design.

The engine, still an upright single-cylinder dohc affair, had new bore and stroke dimensions of 65 mm and 75 mm respectively, producing a total capacity of 248.8 cc. The cams were driven by a redesigned gear train, hiding behind a cover on the right hand side. These gears also governed the Marelli magneto that sat in front of the crankcase and the oil pump mounted on the timing case. The hairpin valve springs were exposed but the remainder of the valve gear was fully enclosed. The oil feed lines to the head were cast into the alloy cover of the gear train, although numerous oil pipes were visible.

In the approved Guzzi mode, permitting a rigid crankcase, there was a hefty outside flywheel, although it was fitted with a light alloy cover that also embraced the primary transmission train. There was a ribbed crankcase, as it had been designed to carry the oil, but in the event a bulky oil tank, with a cooler mounted on the front, was placed under the saddle.

The new 250 cc dohc model *(Sauro Rossi)*.

The quarter-litre engine in its first incarnation *(Mick Woollett)*.

THE PRE-WAR YEARS

Other features were a substantial carburettor and a single exhaust pipe running along the left hand side of the machine to the rear spindle, terminating with a short trumpet. Finally, there was a new four speed gearbox, of Benelli's own design, bolted astern. This package, producing perhaps 26 bhp at 8,400 rpm in its original form, took the machine to a top speed in the region of 115 mph. A penalty for the high state of tune was that the power did not come in until 6,000 rpm.

The cycle parts were unsophisticated. Featured were a single downtube and a simple cradle frame with a swinging-arm rear suspension controlled by plunger spring boxes and friction damping, still some way behind the advanced systems that had already emerged from both Guzzi and Gilera. Similarly, the Benelli's front forks were less than state of the art; they were single spring girders with friction dampers, whereas BMW was already well advanced with its use of telescopics. The tyre measurements were 2.75 x 21 in.

Armed with this potent missile, the Benelli brothers set about recruiting a top notch team. Old boy Rossetti was signed up for yet another term and Nino Martelli and Emilio Soprani joined him. The season progressed smoothly, for although Nello Pagani aboard the perpetually rapid Guzzi collected the important national title, Rossetti and Martelli took the laurels at Bologna and Faenza respectively. However, it was to be at the Italian GP that the 250 cc Benelli was to mark its potential in the most convincing of styles.

For 1938 the format of the European championship had, at long last, been revised with points being awarded at eight major Grands Prix. The DKW team had well and truly monopolised the 250 cc stakes, with Kluge wrapping up the title. But at Monza, Soprani registered a famous victory, less than a second ahead of his colleague Martelli. Indeed, Benelli's cup was overflowing because Rossetti, having pitted for a plug change, hauled back his lost time to such good effect that he caught Guzzi's Tenni on the line, cementing a resounding one-two-three for the Pesaro marque.

This stunning Monza performance was, inadvertently, the prelude to the one major success that had so far eluded Benelli - a Tourist Trophy. The 250 cc race at the Italian GP had been run concurrently with the 350 cc event and the first two Benellis came home at a speed faster than that attained by the Junior victor, the European champion Ted Mellors aboard his Velocette.

Mellors competed in his first major race, the Amateur TT - the precursor of the Manx GP - in 1927. Classic succes came his way three years later when a 250 cc New Imperial carried him to the French GP. In all, between 1930 and 1939, Mellors notched no less than 20 Grand Prix wins.

The revised racer of 1938 *(Mick Woollett)*.

Despite this enviable run, Mellors nurtured one unfulfilled ambition: a TT victory. Having been blown into the weeds by the Benellis in the Italian GP, Mellors studied them in the pits at Monza after the race and planted the idea of an entry in the following year's TT. And so it was that, for the 1939 series in the Isle of Man, Mimo Benelli took along a solitary race bike, a few spares and a single mechanic, Magnani, dedicated to providing Ted Mellors with his sought-after Trophy.

The Italian press was not, initially, over-impressed with this seemingly half-baked effort. One scribe wrote: "Giuseppe is the designer and creator of the machine; Giovanni is the builder and he refines and tunes it to perfection. Although it is true that at a busy period for the factory neither Giuseppe nor Giovanni has much time to waste racing, it is also true that an event such as the TT should have been assigned to one of the technical team if it had been intended to give the effort the full support that such a serious race requires. Mimo Benelli, representing the factory, works in the company in a commercial role. The TT is certainly an event of great commercial significance but the preparation of a machine and the planning requires a full race department. By sending Mimo to Douglas, the factory has given the entry official works status without providing the mechanical back-up that it normally gives to a race of this importance."

Nevertheless, Mellors had reason to be confident. By now the bike was producing 27 bhp at 9,500 rpm and, crucially for a jaunt over the demanding Mountain circuit, was reliable. Reportedly, Magnani's sole task once on the Island was to change the jetting.

In the dry, Mellors established the fastest pratice lap at over 80 mph, pipping Kluge's existing official record. However, it was to be no one horse race, for the assembled opposition was formidable. Guzzi sent over a couple of supercharged singles for Woods and Tenni and the DKWs mounted a notable challenge with Kluge, Wunsche and Ernie Thomas who were actually to finish second, fifth and eighth respectively.

At the end of the first lap Tenni led his stable-mate Woods by 8 seconds with their Benelli rival another half a minute astern. But on the next tour, Mellors moved into a lead which he was never to relinquish, as Tenni suffered plug trouble and Woods slowed. The Irishman recovered to set the fastest lap of the race on the fourth round but he was way behind the Benelli. Indeed, he had expected that Mellors would win, explaining that: " The Benelli was not as rapid as the Guzzi or the DKW but, not being supercharged, it was more reliable and Mellors was very experienced over the course."

Once the race was over, the victorious machine was test ridden by Graham Walker who had by then become the editor of "Motor Cycling". As no rev counter was fitted, Ted Mellors told him to "rev it until the exhaust note doesn't alter." Walker reported that he had taken it to about 110 mph and found the engine to be "unburstable". Meanwhile, "Motociclismo's" reporter Silvio Giacotti relayed news of the victory to Italian fans, and unfairly played down the role of Ted Mellors with the words, "Frankly, he is not Tenni, Woods or Kluge."

A second Benelli had been entered in the Lightweight TT, ridden by a Belgian who appeared on the entry card as G.V.Dickwell, although his name was actually Barbe. In 1938 he had acquired one of the old rigid framed 250 cc racers from the factory and he campaigned it throughout Europe for two seasons. Indeed, after the War, the machine was to have an extensive and fascinating race history.

Meanwhile, back in their homeland, the Benelli racers performed creditably. Rossetti emerged as national 250 cc champion and took the honours in the Milano-Taranto open roads race, that had for some years been dubbed the Coppa Mussolini in honour of Il Duce, while Martelli won at Foggia.

The Italian GP was cancelled as most of the would-be competitors became embroiled in another more serious conflict during the first week of September. However, Mussolini, anxious not to leap to Hitler's aid too precipitously and thus opt for the losing side, refrained from plunging his nation into armed conflict. Hence, the Italian manufacturers all prepared enthusiastically for the 1940 domestic season. The existing Benelli, although basically a little dated, was by no means yet outclassed by its rivals. It was probably as light as the Guzzi and was reputedly better handling on the twisty courses but it was difficult to ride as it had a narrow useful power band. More to the point, it was, however, in danger of being left behind in the power stakes as the Guzzi technicians perfected their supercharged single.

As a long term solution, Giovanni Benelli proposed a supercharged four cylinder racer but since 1938, as an interim measure, he had been toying with tacking a supercharger onto the existing single-cylinder steed. The blower was fitted to the flywheel cover. The almost mandatory compensating "lung" (as the Italians quaintly called it) or reservoir was placed behind the cylinder; it was aircooled with extensive circular finning. Thus boosted, power was now up to 35 bhp at 9,000 rpm and, by dint of attention to detail, weight was reasonable at 245 lb. The machine had a solitary outing during the truncated 1940 season when Emilio Soprani rode to victory at Verona. Sadly, in June of that year Mussolini plunged his country into the War during the course of which Pesaro was occupied by German forces. Legend has it that during that occupation one of the German soldiers, anxious to sample the delights of the blown single, was caught unawares by its phenomenal acceleration, riding it into a wall and wrecking it. Despite this mayhem, the bike emerged from the War and was rescued from oblivion in Pesaro in the 1950s.

Although scarcely raced, and indeed similarly remembered, Benelli's first 500 cc racing venture merits a brief description. Like its 250 cc sister, it was an unspectacular design, sparing of invention, being based on the tried and trusty 175 cc model. It was prepared in readiness for the 1935 season.

The vertical single cylinder featured bore and stroke dimensions of 85 x 87 mm, producing 494 cc, and as would be expected it boasted a twin cam head, governed by a gear train on the right hand side. The bike did however sport one or two distinctive features. The cover over the gear train was a much more substantial matter than that on the racer's smaller sister, presenting all in all a much cleaner image. The magneto was still at the base of the cylinder but this time sat behind it, while the flywheel was internal and there was only one exhaust pipe, on the right.

Ted Mellors and his mechanic Magnani in the paddock behind the TT grandstand, 1939 *(S.R. Keig)*.

Mellors at Governor's Bridge, 1939 TT *(S.R.Keig)*.

The supercharged 250 cc model *(Sauro Rossi)*.

The 1935 500 cc machine *(Sauro Rossi)*.

THE PRE-WAR YEARS

The starting grid at the 1935 Monza GP. Sandri, number 30, is in the centre of the photograph with the works Benelli *(Benelli)*.

The cycle parts were virtually carbon copies of those on the 250 cc racer. The motor developed about 50 bhp and propelled the 300lb. machine to a top speed of approximately 125 mph. These statistics hardly inspired confidence. The harsh truth was that the racer was outdated even before it took to the tracks, bearing in mind that the CNA Rondine enjoyed 60 bhp and a top speed of more than 135 mph while the twin cylinder Guzzi was already being lined up for the benefits of rear springing.

Nevertheless, the machine was carted to Tripoli for the GP at the Mellaha circuit, which owed its name to the salt lake which it encircled. The 8 mile circuit, being little more than a loop, was one of the fastest in racing and enjoyed some of the finest facilities, thanks largely to the proceeds of the infamous lotteries associated with the car GPs.
Recently erected was the world's finest grandstand, with a huge cantilevered roof providing much-needed shade for 10,000 spectators.
The new pits complex was similarly the last word, being built in dazzling white concrete and dominated by a huge sparkling white timing tower. The surrounding plentiful palms, lush green lawns and colourful flower beds provided an air of sophistication.

With the delights of a casino and a theatre to hand, races at the circuit were hugely popular with riders and spectators alike, as they represented the social highlights of the year for the Protectorate. Dick Seaman, the English star of the all-conquering Mercedes equipe, once dubbed the event as "the Ascot of motor racing."

Yet side by side with such grandeur existed the realities of an impoverished society. For instance, during practice the riders encountered a number of novel obstacles: peasants walking along the road carrying their wares to market and countless stray camels and donkeys wandering untented across the track.

The GP, on 31st March, was the first major event of the 1935 season and all the teams were out to impress. As well as a lone Benelli, the CNA four cylinder racers were on show for the first time in the hands of Piero Taruffi and Amilcare Rossetti while a handful of Guzzi vee-twins were entered. The CNA duo romped home convincingly, leaving the remnants of the Guzzi squad in their wake. The solitary Benelli sadly failed to last the distance.

Selected to ride the Benelli was arguably the most fascinating and charismatic of all pre-War riders: Guglielmo Sandri. Born in 1906, Sandri had been a top-notch racing cyclist and tennis player in his youth. Like many of his contemporaries, he combined professional life on two wheels with a career on four wheels and he had driven for the Maserati squad. As a motorcyclist, his best years were still to come, being those immediately before the War when as a member of the Moto Guzzi squad he took a hatful of prestigious races and world records.

But perhaps the most extraordinary aspect of the man was that he combined his life of speed with a dedication to the arts. He was particularly devoted to opera and on occasion he would hasten from the track to sing in public performances.

Sandri was back within the Benelli ranks a few weeks later for the Monza GP - which the foreign teams were boycotting. On this occasion the 500 cc Benelli did at least last the distance but it was once again acutely disappointing, being noticeably the slowest of the works entries.

The Benelli was also taken to the Lario circuit for the TT, only to be retired by Panella when in a distant fourth spot, whereafter it was not to be seen again in 1935. As late as 1939 there were plans to resuscitate the machine, and the factory could well have experimented with plunger rear suspension and a revised gearbox. In butchered form the bike could have survived the War and parts could exist today but, with the passage of time, "the entire truth and nothing but the truth" regarding the subsequent history and whereabouts of this unique racer is incapable of being established.

If the barely-remembered 1935 500 cc single cylinder bike was one of the least technically interesting and most disappointing machines to emerge from the Pesaro factory, four years later appeared arguably one of the most sensational pieces of racing hardware to see the light of day, anywhere, at any time: the supercharged four cylinder 250 cc racer.

Despite the TT laurels in 1939, Giuseppe and Giovanni appreciated that the days of the normally aspirated engine were over and that supercharging was henceforth a sine qua non. The brothers also realised that single-cylinder engines were not best suited to supercharging. The problems associated with supercharging (the size of the reservoir, cooling and throttle lag) could be considerably reduced, although not entirely eliminated, by the use of a multi-cylinder engine, as the blower's delivery could be readily harnessed to the more frequent inlet strokes.

Benelli unveiled their new model at the Milan Show late in 1939. The cycle parts of the outdated single were plagiarised; the same elementary frame, girder front forks, plunger rear suspension, 2.75 x 21 in. tyres and so on were all borrowed without significant amendment. The only apparent difference was that the tank boasted a badge of four proud lions beneath the name 'Benelli'.

But the engine was a revelation. Each cylinder was of 42 mm bore by 45 mm stroke for a total capacity of 249 cc. In the mode established by the 500 cc Gilera, the four cylinders were mounted across the frame, and were inclined at 15 degrees. The two valves per cylinder were governed by a dohc assembly with a shaped cascade of gears on the right hand side. An extension ran to the front of the engine to drive the magneto. Watercooling was employed with the radiator mounted on the front downtube, while the oil tank was beneath the saddle.

The gear driven blower was a Cozette vane type mounted above the four-speed gearbox. The supercharger employed one carburettor and a compensating lung with generous circular finning. The four exhaust pipes merged into two, with one pipe on each side.

This fearsome power plant supposedly offered an undreamt-of 52 bhp at 10,000 rpm, producing a purported top speed of 145 mph in race trim and 155 mph in streamlined guise for record attempts. Regrettably, the machine never had the opportunity to show its paces on

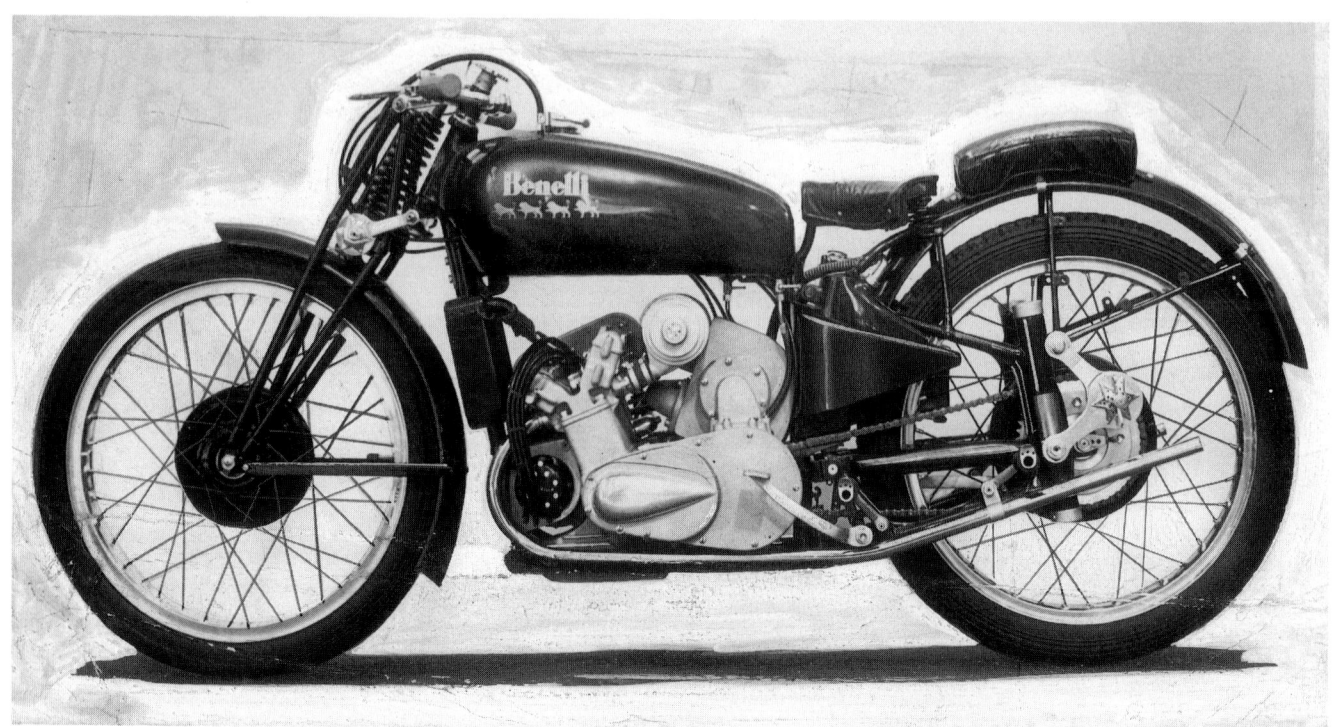

The legendary supercharged four-cylinder 250 cc racer (*Mick Woollett*).

THE PRE-WAR YEARS

A post-War shot of the mighty blown steed. Note the revised radiator and tank.
The Belgian rider Leon Martin and Dario Ambrosini *(Mick Woollett)*.

Factory tester Salvatore Baronciani with the revised blown racer*(Gianni Perrone)*.

THE PRE-WAR YEARS

A 250 cc Monotubo sports model in 1950 action at Finale Emilia; rider Gardo Niccolai *(Gianni Perrone)*.

A 1938 250 cc Monotubo.

The 250 cc Monotubo engine.

THE PRE-WAR YEARS

the race track. An example survived the devastation inflicted on the Pesaro factory by the retreating German armies but, thanks to the FIM ban on supercharging imposed after the War, it was destined never to turn a wheel in anger.

Although not genuine racers, Benelli did produce two more steeds for the track. In 1937 sports machines of 250 cc and 500 cc were added to the factory's range available to the public for road use. They boasted single cam heads, as usual governed by gears. The engines were slightly inclined forwards and featured wet sumps, while the frames were simple open affairs. The smaller model actually shared the engine dimensions of the first generation 250 cc racer, at 67 x 70 mm, but the larger bike, at 85 x 87 mm, owed nothing to the virtually still-born 1935 race model.

When in 1938 the Italian motorcycling federation introduced a new racing formula for sports bikes, Benelli seized the opportunity to offer a limited run of their production bikes in race trim. The bikes were stripped of their surplus accessories, fitted with plunger rear springing and officially designated 'Super Sports' models. As there was only one exhaust pipe, both bikes became known as 'Monotubo' racers.

The nearest the factory came to running a works Monotubo was the support given to Massimo Pasolini from nearby Rimini. Pasolini acquired his 250 cc model in 1938 and between weekend races his bike would be fettled free of charge in the race department by chief mechanic Salvatore Baronciani.

These off-the-peg racers performed creditably in the hands of privateers in the new formula. The Lightweight version was particularly successful and carried Gianni Leoni to a junior category championship in 1939, as well as a class victory in the final edition of the Lario classic in the same year. Jader Ruggeri, son of the star of the 1920s Amadeo, also took one to the laurels for his class in the last of the Milano-Taranto open roads races before the War in 1940.

Indeed, Monotubo machines continued to appear on the starting grids in the immediate post-War years until they were finally outclassed by a new generation of works racers with the onset of the 1950s.

A 250 cc Monotubo in action at Castiglione del Lago, 1994

Chapter Three: The Ambrosini Years: 1945 - 1952.

During the War, the factory was rudderless, as the five brothers were dispersed. Giuseppe joined the army, as did Giovanni who was a chief technician responsible for anti-aircraft defences and Francesco who was employed as a driver. Filippo became a pilot in the air force. Despite their absence, the company prospered initially and for brief spell enjoyed unprecedented prosperity, producing high precision work for the aircraft industry, as well as engines for the OM and Alfa Romeo concerns.

But from the dizzy heights of 1940 there befell the company a series of devastating blows. The Italians' so-called German comrades in arms were not above blatant pilfering, with complete machines being transported back to the Fatherland. Towards the end of the War the Allies bombarded Pesaro heavily because it formed part of the German Army's Gothic line of defence and the factory premises came in for their share of the shelling.

As if this were not sufficient, the partisans did their best to destroy the factory to prevent the Germans making use of it and, in turn, when the Germans pulled out of the Adriatic resort in September 1944, they looted the factory comprehensively and destroyed what was left. So it was that the brothers returned home to a scene of desolation.

Indeed, the mild-mannered Giuseppe was heart-broken by the wholesale destruction of the family's enterprise and suffered a psychological blow from which he never fully recovered.

Nevertheless, under the direction of Giovanni and Filippo, who was the family's presiding financial genius and the company's commercial director, the brothers set about reconstructing their business empire. No stone was left unturned. They borrowed pre-War machines and spares from their workers, they bought a thousand bikes from the occupying American forces, mainly Matchless and BSA machines with some Harley-Davidsons thrown in, and then set about converting them by adding a new rear suspension that Giuseppe had designed. It was a first step back to production.

Of course, the theft or destruction of the racing hardware, the loss of the factory's machinery and indeed the family's inevitable resultant impecuniousity did not permit the brothers to indulge in any road racing activity.

For the most part, the Benelli name was kept to the fore by privateers aboard the 250 cc Monotubo models, such as Bruno Francisci who gave the marque its first notable post-War success at the Tortoreto street circuit in 1945, while in Belgium the ex-Dickwell rigid framed works racer found its way into the hands of Leon Martin. He took it to Pesaro for renovation at one stage but as there were no racing frames available it was fitted with the rear end of one of the sports production machines.

Luigi Ciai, Reggio-Emilia, 1947 *(Sauro Rossi)*.

THE AMBROSINI YEARS: 1945-1952

Dario Ambrosini *(Gianni Perrone)*.

Ambrosini jumps into the lead at San Remo *(Gianni Perrone)*.

By 1947 the factory had assembled sufficient bits and pieces to provide a couple of 250 cc bikes on a strictly unofficial and distinctly half-baked basis. The fortunate recipient of this bounty was a promising newcomer, Luigi Ciai, a second or junior category rider who collected the laurels in races at Turin and Rome.

Highlight of the year, as ever, was the GP which, as the traditional Monza parkland track was in a state of considerable disrepair, was held as a one-off over the Fiera Campionaria street circuit in the centre of Milan. Ciai was outdistanced and finished in fourth spot behind a trio of Guzzi Albatros riders, namely that year's senior category champion Nino Martelli, the junior title holder Alfredo Milani and a youthful Bruno Ruffo. Languishing in fifth, and lapped, was Benelli's second recruit, the familiar Amilcare Rossetti.

Thus far, the Benelli factory had been playing a distinctly secondary role in the revival of Italian road racing but it was unexpectedly thrust into centre stage as from December 1947.

Dario Ambrosini was born in Cesena, not too far to the north of Rimini, on 7 March 1918, into a family of distinctly modest means. His lifelong friend and racing manager Bruno Zoffoli described his initial forays into motorcycling: "Even as boys we had a passion for motorcycles and used to borrow our fathers' bikes for local trips. By about 1937 Dario had obtained a 250 cc Benelli, in virtually a standard road-going state, and began racing, with his first event at Ferrara I think. His first victory came at Verona in 1939 in a third category championship race."

His delayed breakthrough into prominence came in 1946 aboard a 250 cc Guzzi, with the laurels at Pinerolo, Riccione and, most notably, over his home town circuit at Cesena, where he beat Ciai who was campaigning the semi-works pre-War Benelli He confirmed his status right at the pinnacle of his profession when he beat the prevailing genius Omobono Tenni at Mantua later in the year. So impressed was the Black Devil that the master offered to accompany Ambrosini on his sorties abroad in order to refine the apprentice's skills and thenceforth the two became firm friends.

In 1947, Ambrosini repaid his mentor's faith, being crowned as the 250 cc Italian champion. His education continued over foreign street circuits and he notched up victories in Zurich and Lugano. Tenni's claim that Ambrosini was 'the complete racer' did not appear to be too far from the truth.

So this was the shy, modest, good natured man who, in December 1947, seemed to have the motorcycling world at his feet. However, one morning he picked up a newspaper which confirmed his worst fears: the Guzzi company was cutting back on its official racing participation. On an impulse, Ambrosini took himself off to Pesaro, despite the knowledge that the only race hardware in the Benelli factory was the couple of tired ten year old models used by Ciai and Rossetti. Although sceptical at first, the Benelli brothers were impressed by Ambrosini's evident enthusiasm and, having consulted Filippo, who was the most businesslike of the brood, the decision was taken that the factory should mount an official return to the tracks.

Ambrosini pictured at a damp San Remo, with the rear suspension suitably protected from the elements (*Gianni Perrone*).

THE AMBROSINI YEARS: 1945-1952

Umberto Masetti on the grid at Monza, 1949, with his father to his left and Ambrosini behind them *(Gianni Perrone)*.

Masetti riding to third place in the Italian Grand Prix *(Gianni Perrone)*.

The announcement breathed new life into the company and its employees. Giovanni now threw himself energetically into an examination of the racing engines, while Mimo's task was to recover the missing machines. After months of enquiries and travels throughout Italy and Germany, Mimo repatriated various bits and pieces that had been dispersed during the hostilities. Rejuvenated by the prospects, the racing department at Pesaro set to work with a will, producing new frames to house the engines that enjoyed the advantage of new crankcases and pistons commissioned by Giovanni from a foundry in Turin.

Ambrosini's effort was still far from a fully fledged works squad of the sort that virtually every self-respecting Italian company was to field during the mid-1950s. He relied very heavily on his manager, Bruno Zoffoli, and the financial and moral support of another long-standing friend from Cesena, Natalino Bersani. Nevertheless, Ambrosini single-mindedly dedicated himself to the task of perfecting his machinery. From morning to night, he would test the Benelli racers along a stretch of road between Pesaro and Urbino, and he and his tiny band would strip each bike down at least once a week and rebuild it with meticulous care.

As fate would have it, the first race of the new campaign was to be held over the Cesena street circuit, and the Benelli squad set up its headquarters in Ambrosini's house, using the living room as their workshop. As practice commenced, the Guzzi team's manager, Cesare Battistini, immediately produced a brand new 250 cc racer for Ambrosini's use in practice. His hope was that, having sampled the machine, the Benelli teamster would opt for the faster Mandello product.

However, Battistini had failed to take account of Ambrosini's loyalty and the unexpected pace of the refurbished Benelli that enabled the home town boy to top the practice leaderboard. In fairy-tale fashion, Ambrosini took the chequered flag to the delight of his townsfolk, and in front of the embarassed and irked Guzzi overlord Giorgio Parodi. To rub salt into the Guzzi wounds, Ciai brought his Benelli home in second place.

In the middle of May, Ambrosini took the honours in the Swiss GP held over the Circuit des Nations street track just outside Geneva. Throughout the remainder of the season, he collected victories at Monza, San Remo, Lugano and Lecco.

The solitary blot on the horizon came at the all-important Italian GP in September. With the traditional Monza venue still some weeks from a complete restoration, the highlight of the domestic season was held over a 2-mile street circuit in Faenza. Despite a sick motor, Ambrosini scooped second spot, but he was overwhelmed by Bruno Ruffo aboard a works Guzzi, while Rossetti brought his Benelli home in third position. The remaining member of the Pesaro contingent, Luigi Ciai, tumbled at three-quarter distance.

Ambrosini with Giovanni Benelli, holding his jacket, to the left *(Gianni Perrone)*.

THE AMBROSINI YEARS: 1945-1952

A 250 cc field gets underway. Included are Luigi Ciai (Benelli, 14), Fergus Anderson (Guzzi, 19), Ambrosini (sporting his customary number 23) and Bruno Ruffo (4, Guzzi) (*Gianni Perrone*).

Ambrosini at Ferrara; note the rear wheel disc (*Gianni Perrone*).

THE AMBROSINI YEARS: 1945-1952

Bruno Francisci, part-time Benelli rider, but better known for his three Milano-Taranto wins aboard Guzzi and Gilera machinery *(Gianni Perrone)*.

On a personal level, Ambrosini had suffered a devastating blow when his friend and mentor Tommaso Tenni - better known since his childhood as Omobono, for an obscure reason unknown even to his family - was killed in a practice accident at Berne's Bremgarten track in July.

As the 1949 season approached, Ambrosini had three ambitions: the Italian title, a TT success to emulate the great Tenni and the newly instituted world championship that was to be held over four rounds, in the Isle of Man, at Berne, Clady and Monza.

The Benelli factory produced a revised machine which featured slightly more substantial finning on the cylinder, a larger petrol tank, a long racing saddle and a disc over the rear wheel for fast circuits. With the improved fuel that was now available, power was up to 26 bhp at 9,000 rpm.

Reliability was a problem as the season got underway so that the Italian title was as good as surrendered to Guzzi's Bruno Ruffo. Foreign travel brought about a welcome upturn in Ambrosini's fortunes with successes at Floreffe, Voghera and Olten, but these were merely preliminaries to his attempt on the Lightweight TT.

Quite apart from the fact that the Manx races represented the first round in the world championship series, Ambrosini nurtured a burning ambition to capture a Tourist Trophy race. Appreciating the scale of the task before him, he approached the problem with his customary thoroughness. He arrived in the Isle of Man twenty days before the race and immediately set about familiarising himself with the daunting Manx roads. As well as putting in as many practice laps as possible, he pinned a huge map of the course on the wall opposite his bed and would spend hours gazing at the landmarks, such as Quarter Bridge, Ballacraine, Ramsey Hairpin and so on.

The 250 cc race began, in those days, with a massed start and for the first lap Ambrosini was content to shadow the assembled ranks of the Guzzi riders who constituted his major opposition: Lorenzetti, Barrington, Tommy Wood and the youthful Dickie Dale. As Ambrosini gingerly negotiated the circuit's slowest corner, Governor's Bridge, in a respectable fifth spot, he dropped the red Benelli, putting himself out of contention and injuring an arm in the process. Needless to relate, one of the Guzzi hordes, Manliff Barrington, won the race.

Ambrosini was devastated by this unaccustomed setback and he carried his mood of depression to the Swiss GP at Berne some days later. Riding as a shadow of his former self, he was judged to have been fortunate to salvage second spot 40 seconds behind Bruno

Ambrosini and Francisci in team formation *(Gianni Perrone)*.

Ruffo's Guzzi. His championship hopes were dashed when he wrecked his bike in a spill in the Lugano international, forcing him to give the Ulster round a miss. Although Cann took the spoils, Ruffo secured the points for second place.

Ambrosini had thus far ridden the season without a team mate, as old stager Rossetti had retired and Luigi Ciai had departed for Parilla. The Benelli management decided to rectify this situation for the Monza showdown and hoped to recruit Dickie Dale but, when that proposal fell through, Benelli opted for the latest Italian sensation Umberto Masetti, the newly-crowned Italian 125 cc champion aboard a Morini.

Although fully expecting to blow his motor, Ambrosini simply cleared off and after an hour's riding won the Italian GP more than half a minute ahead of the Mandello factory's leading rider, Gianni Leoni. Masetti made an appalling start and was in eleventh spot at the end of the first lap but came through to register third place. However, the canny Ruffo rode a carefully judged race to finish fourth, which was sufficient to secure the title.

The Benelli star was destined to finish the season virtually empty handed, for although he won the final Italian title race at Mantua, backed up by Masetti, Ruffo once again rode tactically to take third place and thereby clinch the domestic championship.

If 1949 had been an unmitigated disappointment, 1950 witnessed the fruition of Ambrosini's genius. A retirement at San Remo when, in wretched conditions, he twice shunted the bales and a victory at Olten were but the precursors to his sortie to the Isle of Man for the first of the year's world title races.

An attempt by Dickie Dale's manager Sam Coupland to secure a Benelli for his rider fell through when the ACU refused to accept a last minute entry and so once again Ambrosini faced the Mountain circuit as a lone wolf. He was however accompanied not only by his faithful mechanic Maroccini and Mimo Benelli but also by Bruno Zoffoli, and they planned the campaign meticulously.

Once again, Ambrosini packed in as many laps as he could manage during practice to acquaint himself with the circuit. Furthermore, auxiliary tanks were fitted, one atop the main tank and another built into the rear mudguard assembly. Zoffoli described the team's other trick: "Because the circuit was about 60 kilometres in length, if a rider relied on signalling from the grandstand his information would always be effectively out of date, and so we decided to set up a signalling station at that famous bridge, Ballaugh. I was stationed there during the race and once the riders came through I used a public telephone to get through to the grandstand so that I could give Maroccini

THE AMBROSINI YEARS: 1945-1952

Ambrosini, mechanic Maroccini and George Wilson of 'The Motor Cycle' in the Isle of Man, 1950 (*Mick Woollett*).

In the TT winner's enclosure, 1950, after an epic encounter, Maurice Cann, Count Lurani, Ambrosini and Maroccini (*Mick Woollett*).

THE AMBROSINI YEARS: 1945-1952

The Italian GP, 1950. The Benelli squad included Ambrosini (52), Miele (64) and Piergiovanni (92) *(Gianni Perrone)*.

Piergiovanni before his retirement, Monza 1950 *(Gianni Perrone)*.

Miele gets down to it at Monza *(Gianni Perrone)*.

Ambrosini and Maroccini celebrate their Monza GP success, 1950 (*Gianni Perrone*).

THE AMBROSINI YEARS: 1945-1952

The revised 1951 machine with updated frame and front and rear suspension (*Mick Woollett*).

The Italian racing fraternity at play. Seated: first from the left is Bruno Francisci, third is Tenni and sixth is Balzarotti. Amilcare Rossetti stands next to Ambrosini.

Just before the fateful accident at Albi: Orlando Valdinoci, in his vest, in conversation with Ambrosini, with Natalino Bersani in the dark shirt.

THE AMBROSINI YEARS: 1945-1952

A sad tribute at the scene of the world champion's fatal crash.

the latest information, and so we were only half a lap behind with our signalling." Given the nature of Ambrosini's race strategy this ploy was to be crucial.

Ambrosini completed the first of the seven laps in a remote fifth place, a minute behind Cann and Wood who were vying for the lead. The Benelli rider was mindful of the allegation that in the previous year's event he had been over-impetuous and he subsequently reported that in addition the bike's handling had been hampered by the extra tank over the rear wheel. As the race progressed, he consistently whittled down his lap times, so that from a first lap of 30 minutes 50 seconds his last was a shade under 28 minutes. He gradually picked up the lost time on Maurice Cann and started the final tour fifteen seconds in arrears. He finally caught him in the region of Creg ny Baa, a couple of miles from the chequered flag and so it was that Ambrosini rode to victory but a few yards ahead of his Guzzi adversary, to become only the second Italian to capture a Tourist Trophy.

He reported after the race that he had increased the tempo as planned, revving to 9,500 rpm and facing no problems once the fuel load had lightened. At the Villa Marina prizegiving he was embarrassed as he spoke no English but Count Lurani, a noted Italian motor sport fan and author, prompted him into saying "I am very happy." He returned home to Cesena and a hero's welcome.

From then on, Ambrosini had a clear run to his cherished world title. He beat his greatest rival, Bruno Ruffo, by fifty seconds to collect the Swiss GP at the Circuit des Nations in Geneva. Dickie Dale brought the second Benelli home just astern of the Guzzi.

The season's third classic, the Ulster, saw Cann ride to a comfortable victory as Ambrosini was content to collect the points for runner-up spot. He had taken three bikes to the Clady circuit. In practice, a conrod went clean through the crankcase on one machine and another motor seized, while during the race the third engine suffered from faulty carburation. The bike that had seized was lent to Rudge diehard Roland Pike for the race during which the engine seized yet again, no less than five times, once in mid-air and finally on the notorious Clady straight.

Ambrosini cemented his title with an unchallenged ride to the chequered flag in the Italian GP at Monza. The Benelli factory had entered an unusually large squad for the all-important event and was rewarded as Bruno Francisci picked up third place. The other two bikes were plagued by plug trouble, so that Miele finished a lap down in eighth and Piergiovanni retired.

The Italian and world champion simply continued in unstoppable form as 1951 got underway, with victories at San Remo, Floreffe, Codogno and the Swiss GP at the formidable Bremgarten Forest circuit.

From Berne the Benelli squad headed straight for the Island for the second title round. The race featured a staggered start but this time, over four laps, Tommy Wood pipped Ambrosini by 8 seconds. Interviewed after the race, Ambrosini, who was described over the air waves as a commercial farmer and main distributor for Benelli, gave a radio interview, transmitted to Italy, in which he explained that although his engine was faster than that used in 1950 it had been bedevilled by over rich carburation, so that all in all he was not too disappointed with second place.

After a visit to a number of motorcycle factories in Birmingham and Coventry, Ambrosini returned for a few days to Pesaro and thence to the fateful French GP at Albi. Although Ambrosini's race hardware was effectively a pre-War device it had been the subject of continuous development and Bruno Zoffoli was confident that by 1951 something like 33 bhp was available. Additonally, the factory had come up with a new frame, featuring duplex bottom tubes and telescopic suspension to replace the outdated girder fork and plunger rear springing that had been in use thus far.

Unusually, Ambrosini was accompanied to Albi by quite a team and on the day before practising began Natalino Bersani drove both Ambrosini and his new team-mate Orlando Valdinoci around the 5-mile triangular Raymond Sommer circuit. As they travelled through the hamlet of Bernadie, they discussed the appropriate gear for a particularly rapid curve. Valdinoci was all for going down one gear but Ambrosini assured him that it could be taken in top.

Tragically, during official practice, Ambrosini came off at this very spot and hit a telegraph pole. He was dead when Gianni Leoni stopped at the scene of the accident, and with him stopped the heart of Benelli's racing enterprise. The cause of the accident was never satisfactorily explained. The popular theory was that he had skidded on melted tar brought about by a heatwave, but Bersani attributed the loss of control to the inadequacies of the largely untried frame. What was certain was that Italy had lost its favourite racing motorcyclist.

Devastated by the tragedy, Benelli withdrew from competition for the remainder of the year, although Masetti was entrusted with bikes for unofficial practising at Monza just before the GP. The indications were that they would not be as rapid as the inevitable Guzzi opposition and so the bikes were whisked back to Pesaro.

Over the winter the racer underwent further refinements, acquiring 18 in. wheels instead of the 19 in. wheels of the previous season, a larger front brake and a five speed gearbox, which had been proposed by Zoffoli before Ambrosini's death. The new bike, featuring on this occasion an experimental four valve head, was given its debut at Cesena by Alano Montanari and he brought it home in a respectable third place behind Lorenzetti and Ruffo.

Thus encouraged, the team started out on the world championship trail, turning up at Berne with two machines that were entrusted to Luigi Ciai, back in the fold after his excursion to Parilla, and none other than Les Graham, who had entered the 250 cc race on a Velocette that failed to turn up. The Benellis had been revised further, and they now employed a very large petrol tank that was prolonged to and embraced the fork head in the aerodynamic mode that was fashionable at the time.

Graham, the 500 cc champion in 1949 aboard an AJS, challenged Fergus Anderson and Lorenzetti initially but the loss of top gear put a stop to that, so that his eventual third place was handsome enough. Ciai made a poor start and did not finish. Duly impressed by Les Graham, Benelli offered him a bike for the TT but, contracted to MV Agusta, he declined the offer.

After the TT, the machines were offered to Bill Lomas and Cecil Sandford who were invited to Pesaro for testing. Having been shown the racing hardware at the factory, including the pre-War four cylinder model, the Englishmen were subsequently taken to the scene of their afternoon's sortie: the long straight outside town that had been Benelli's proving ground for years. Duly forewarned that high-speed exploits were imminent, the local constabulary would obligingly close the road and indicate a detour for the public!

The bikes, riders and mechanics then set off for the Dutch TT but the first practice session indicated that the engines were plagued by carburation problems. In an attempt to resolve matters, new Dell'Orto carburettors were fitted and after an all-night stint by the mechanics Lomas was up before dawn on race morning to test his bike over the public roads near his hotel. Travelling at something estimated at over 110 mph, he spotted a dog meandering across the road and, in taking avoiding action, mounted the grass verge. With that frightening episode, the Benelli team decided that the bikes simply were not competitive and the two English stars were non-starters. The factory's entries for the end of season Italian GP were scrubbed and although Ciai rode the 250 cc racer on a couple of occasions during 1953, without success, the truth was that the career of the celebrated pre-War single effectively died with the great Dario Ambrosini.

THE AMBROSINI YEARS: 1945-1952

Luigi Ciai, astride Graham's bike, practice in the 1952 Swiss GP *(Sauro Rossi)*

The 1952 250 cc Benelli *(Sauro Rossi)*

Chapter Four: The 250 cc Single Reborn

Throughout the 1950s, a brace of single cylinder 250 cc Benelli steeds could be found on the tracks. One was the pre-War model that had been campaigned by the Belgian rider Dickwell, who was actually called Barbe but had assumed his nom de guerre when he began his racing endeavours under the age of 18, contrary to the regulations of the Belgian federation.

After the War, Dickwell's compatriot Leon Martin raced the much-modified Benelli until in turn he sold it to the marque enthusiast and privateer racer Norman Webb. The poor bike had changed hands on a number of occasions and had undergone further conversion by the time it was once again in the hands of Webb, at which point, in 1960, it was analysed by John Griffith for his 'Built for Speed' series in "Motor Cycling." All in all, Griffith concluded that it was a far cry from the bikes that had emerged from the Pesaro factory.

Another Benelli in private hands was what was probably a 1952 version that had passed to favoured privateer Giordano Bon. Duly updated in subsequent years, this machine's only performances of any note whatsoever were at the tail end of the results in the Italian GP in 1959 and 1961.

However, a quarter-litre renaissance was at hand. The Benelli brothers opted to reenter the fray with a single-cylinder 250 cc racer, that was generally, albeit mistakenly, believed to be nothing more than an update of the Ambrosini/Graham model of half a dozen years earlier. That myth took root in English racing lore and can perhaps at least in part be attributed to the race correspondent of "Motor Cycling", Norman Sharpe.

Sharpe undertook a tour of the Italian racing factories to investigate their proposals for 1959 and he informed his readers, lyrically, that "lying in the foothills of the Apennines, just south of the vast plain of the River Po, the small Adriatic town of Pesaro is famous as

The re-born 250 cc single, 1959 *(Sauro Rossi)*

Dale in the Imola Gold Cup, 1959 *(Gianni Perrone)*

The Imola Gold Cup meeting 1961 and Spaggiari (40) leads Grassetti
(Gianni Perrone).

Duke and Grassetti line up for the German GP, 1959 (*Mick Woollett*)

the birthplace of Rossini, composer of 'The Barber of Seville'. And shaving has been one of the chief preoccupations of the Benelli factory during the last few months - shaving pounds and ounces off the Ambrosini racer."

In fact, although the new bike's layout was superficially similar to that of the 1951/52 model, it was to all intents and purposes a different animal. It was the work of a young engineer, Aulo Savelli, who had joined the company soon after obtaining his degree. Savelli was asked to re-cast the ex-Ambrosini machine, and that he did during the course of 1958.

When the racer was unveiled in readiness for the 1959 season, it enjoyed the upright single-cylinder layout of its predecessor but the similarities just about ended there, although the gear driven dohc arrangement was also retained. The motor's dimensions were now 70 x 64.8 mm for 249.2 cc and the press handouts claimed 30 bhp at 10,300 rpm, while Giovanni Benelli was content to hint that 33 bhp was being pushed out. There was provision for either coil or magneto ignition, and an integral six-speed gearbox was incorporated, while primary drive was now by gear.

A conventional duplex tubular frame was employed, with standard telescopic forks. The wheels, originally 19 in. affairs, were stopped by new central drum brakes. Weight was probably of the order of 260 lb., although the Benelli publicity machine did its best to shed another 40 lb.

Giovanni Benelli's major problem was in finding a top-notch rider for the 1959 season. Having tried and failed to induce Libero Liberati and Bill Lomas to sign up, he returned to an old friend of proven ability, Dickie Dale, whose career since his days as a second-string to Ambrosini had included spells on Gilera, MV Agusta, Guzzi and BMW works machinery.

Having undertaken tests at Monza, Dale almost gave the new Benelli a glorious debut in the Imola Gold Cup, the principal early season race in Italy. Dale mixed it with the Morini teamsters Liberati and Emilio Mendogni. Alas, Benelli's ambitions were dashed when two valve springs broke, despite which Dale managed to finish in second spot, a mere 15 seconds behind Mendogni. Grassetti, having been slowed in the preliminary stages, took fifth place on the team's second bike.

The squad next appeared in the German GP with a star rider in the ranks, being none other than Geoff Duke. Duke's NSU Rennmax engined special had blown up in TT practice and, anxious for a 250 cc ride, he took over Dale's entry at Hockenheim. Duke, in sixth, and Grassetti, in eighth, were humbled by winner Ubbiali (MV), and the Benellis were carried back to Pesaro for hasty refinement, including the introduction of a twin plug head. The tank was lengthened and lowered, and also featured a deep cut-away on the right-hand side to facilitate access to the additional plug.

The revised bike carried Duke to third at Kristianstad in the Swedish GP, well beaten by Hocking (MZ) and Ubbiali. Dale was sixth. The world title trail petered out disappointingly, when Duke could do no better than tenth at Monza. A much needed morale boosting victory was, however, collected at the Locarno international in September.

Ernst Degner, on the rapid two-stroke MZ, heads the Benelli duo at Imola, 1961 *(Gianni Perrone)*

Silverstone, April 1962. Redman, works Honda, ahead of Hailwood on Fron Purslow's Benelli *(Nick Nicholls)*.

That was the celebrated occasion on which, almost despite himself, Duke collected a hat-trick of wins over the tortuous street circuit with its tight corners and tram lines. Having already won the 350 cc event, and with the 500 cc class still to come, Duke asked the Benelli management if he could be released from his ride. The Iron Duke was prevailed upon to start and settled into second place behind his team-mate, the meteoric Grassetti. When Grassetti pitched himself off on the circuit's solitary fast corner, Duke had little option but to carry on to the chequered flag, outspeeding Taveri's rapid MZ into the bargain.

Indeed, the former world champion subsequently took the laurels in the blue riband class but the demands and strains of that day's racing were instrumental in prompting Duke to bring his illustrious career to a conclusion.

Over the winter, Savelli and his mechanics devoted considerable efforts to developing the quarter-litre single. Indeed, a desmo version that had been on the cards for some twelve months or so was bench tested, but by the start of the 1960 season basically the same frame and engine were in use. Nevertheless, over 30 lb. had been pared off the racer, a lower, more compact dolphin fairing had been designed and larger brakes had been incorporated.

Once again, attracting quality riders was proving a problem. So it was that a veritable hotch-potch of riders sampled the Benelli single during the 1960 early season races. In the first of the Italian title races, at Cesenatico, Bruno Spaggiari was the Pesaro team's choice. Born in 1933 in Reggio Emilia, Spaggiari's career had thus far been almost exclusively devoted to racing Ducati motorcycles. His initial spell with Benelli was both short lived and unspectacular, as he lagged well behind a trio of speedy MVs at Cesenatico and eventually retired.

A few days later, at the important Imola meeting, languishing in fourth, fifth and sixth were three Benelli singles ridden by Grassetti, Dale and newcomer to the brigade Bob Anderson.

Another to sample the Benelli at Imola was the NSU rider and enthusiast Jack Murgatroyd who found himself in hot water when he was reported as declaring that the Pesaro model was nowhere near as fast as his antique NSU. Murgatroyd denied the tale and was happy to confirm, diplomatically, that, on the contrary, the Benelli was indeed rapid and "handled like a dream, being as smooth as silk." On the downside, he admitted that it was difficult to start and had next to no power low down.

Of course, the Benelli squad was fully occupied in developing the new racer, that transpired to be a four-cylinder 250 cc model, and the outdated single was relegated to one-off outings henceforth.

In 1961, Grassetti and Spaggiari both retired at the traditional early season Cesenatico meeting but then improved to finish on the podium behind Provini (Morini) at Imola. Thus encouraged, they embarked on the world championship trail.

At Montjuich Park, Spaggiari cast himself off twice but Grassetti stayed afloat to take a more than respectable third ahead of Redman's Honda four. Other points scoring performances were turned in by Grassetti at Clermont-Ferrand (5th), Assen (4th) and Monza (6th). The non-championship Adriatic GP, held at Opatija, was the scene of Grassettti's best result of the year when he ran to victory in

At the end of its career, the Benelli is raced by John Cooper, Oulton Park, August 1964 *(Nick Nicholls)*.

THE 250cc SINGLE REBORN

Pictured at Braddan Bridge, Hailwood carries on regardless in the 1962 Lightweight TT (*Nick Nicholls*).

The Hutchinson 100, Ralph Bryans at Silverstone, April 1963 (*Nick Nicholls*).

THE 250cc SINGLE REBORN

July ahead of, it must be admitted, a field of privateers. The year was generally a bleak one for the tiny company, and August brought further sad news with the death of Francesco Benelli in Ancona where for some years he had run his own business.

Although the factory's new multi was race-worthy by 1962, the single was campaigned with some modest fortune both on the continent and in England. The Belgian rider Toussaint took his private bike to a handsome sixth spot in the Spanish GP at Montjuich Park and later in the year the long standing Benelli-Motobi stalwart Paolo Campanelli matched this with his performance in the Italian classic at Monza.

The appearance of the Benelli single in England was thanks to the sterling efforts of Fron Purslow who acquired a model in March. The racer, sporting the wasp-waisted fairing that had been introduced in 1961, was tested at Mallory Park by Derek Minter, Ron Langston and the reigning 250 cc world champion Mike Hailwood.

Hailwood was sufficiently impressed to race the bike a week later, winning the Mallory international. A few days later, on the wide open spaces of the Silverstone airfield circuit, Hailwood was only just pipped by Redman aboard his Japanese multi, but his fortunes imminently suffered a downturn at the Imola Gold Cup meeting when he was compelled to retire with ignition trouble.

Hailwood played a game of musical chairs for the TT. He was initially entered by King's of Oxford, his father Stan's firm, on a twin-cylinder MV, but when this fire-engine did not materialise from Gallarate there were hopes of a works entered Benelli multi. This proved to be equally elusive and so it was that Mike the Bike rode Fron Purslow's Benelli.

At the end of the first lap, Hailwood languished in sixth, but with McIntyre's subsequent retirement, and by upping his pace, Hailwood moved into third place for a couple of laps. As the race unfolded between Honda rebel Minter and his notional team leader Jim Redman, drama unfolded in the pits at the end of the fourth lap when Hailwood stormed in with a loose fairing. It took Fron Purslow 40 seconds to tear it off, whereupon the world champion set off without any semblance of streamlining. Sadly, Hailwood's effort against the Oriental multis came to grief at Quarter Bridge on the final circuit when he coasted to a halt. It was reported that the cover had come off the contact breaker, allowing dirt to clog the points.

Hailwood's exploits with the Benelli single were not quite over, for he rode it in the San Remo international and indeed in the Italian GP, only to retire on both occasions.

A number of top flight riders sampled the Purslow Benelli during this period. For instance, the machine was entrusted to the up and coming young Irishman Ralph Bryans for the 1963 TT. Completing the first lap in ninth spot, the Benelli's motor gave up the ghost at Quarter Bridge.

A second Benelli single was also seen on the English tracks in 1962 in the hands of Alan Dugdale. Dugdale took the bike to the Isle of Man, and completed the first lap of the Lightweight TT in a satisfying eleventh spot, but unfortunately the tank was dry by Guthrie's Memorial on the third tour. In 1964 this machine was sold to the entrant Colin Lyster.

Indeed, the factory retained a single as a practice hack and from time to time it would be wheeled out for race use by Provini, who was signed as team leader in 1964. However, the days of the Benelli single, as a competitive top rate machine, were well and truly numbered. After a career enduring for more than a quarter of a century, it was at long last to be pensioned off.

Chapter Five: Post-War Miscellany

The Leoncino

In the aftermath of the War, the Italian populace craved an economical means of transport and so it was that the era of the cheap moped or ultra-lightweight motorcycle dawned. Possibly the standard bearer of these machines was Ducati's hugely popular Cucciolo - the little pup - originally offered in clip-on engine form and subsequently as a complete bike. Other notable efforts were Garelli's Mosquito and the Guzzino, the baby Guzzi.

To some extent, Benelli missed out on this vast new market. Having been devastated during the conflict, the factory was not immediately capable of bulk capacity production and hence it was not until 1949 that Giovanni's offering for this seemingly insatiable demand emerged: a 98 cc motorcycle dubbed the Letizia, named after his daughter. This model spawned the renowned 125 cc Leoncino - the lion cub - that, in its original two-stroke form, was produced shortly thereafter in both Turismo and Sport models.

The Leoncino's engine enjoyed square dimensions at 54 x 54 mm for 123.6 cc and, with a compression ratio of 7:1, boasted 6.5 bhp available at 6,600 rpm. Other features were a 20 mm Dell'Orto carburettor, a flywheel magneto, gear primary transmission, a multi disc clutch in an oil bath, and a four speed unit construction gearbox.

The frame was a simple spine affair, with the engine-gearbox unit slung underneath, pivoted rear forks and telescopic front forks. The tyres measured 2.50 x 19 in. and there were flimsy drum brakes. The machine weighed 185 lb. and could attain a top speed of 65 mph. The Leoncino began to garner some modest success when put to road race use in 1952. Cristoforo Fattori took a bike to the laurels at Rimini, Spoleto and Forli and thus implanted a few ideas in the mind of Mimo Benelli.

It was this humble machine, barely tweaked for competition purposes, that was to thrust the marque to the forefront of the racing scene and indeed the attention of the sports fan in 1953. That year witnessed the revival of the Giro d'Italia, literally the Tour of Italy, that originally had been one of the classic events of the 1920s.

In April 1953, Bologna's daily sports paper "Stadio" sponsored the first reincarnation of the Giro, an open roads race that was to be regarded as a companion to the other such classic, the Milano-Taranto. But whereas the latter was an energy sapping non-stop blast from the head to the toe of the peninsula, and was open to all classes of machines including genuine GP machinery equipped with lighting sets, the Giro was organised as a stage race, over half a dozen days or so, and was limited to sports production motorcycles of up to 175 cc.

The Benelli brothers recognised the Giro as a golden opportunity to participate in the sport without incurring outrageous expense and also to cull some very welcome press coverage. The chosen Benelli hardware was of course the Leoncino, scarcely revised. The team's hopes were pinned on two newcomers, Paolo Campanelli and Leopoldo Tartarini, and the effort was co-ordinated by the enthusiastic Mimo.

At the end of the race, Tartarini took the chequered flag, with Campanelli in second place, to register a comprehensive if totally unexpected success for the Leoncino. Undoubtedly going overboard, "Stadio" recorded: "Mimo Benelli was present at the scene of victory and, perhaps, his steamed up glasses hid his tears of joy. The 125 cc two-stroke (having left at a disadvantage against the four-

A Benelli Leoncino photographed at the Misano Historic Grand Prix.

POST-WAR MISCELLANY

Left: The 125 cc engine.

Below: Leopoldo Tartarini during the 1955 Giro *(Sauro Rossi)*

Fernando Bruscoli in readiness for an event in Pesaro, 1955.

A Leoncino in the paddock at Cattolica, 1991.

strokes according to many), the glorious Benelli marque and its youthful riders have conquered the most important race of the year."

The rival sports paper "Tuttosport" reported that "Mimo Benelli was so overwhelmed that he could not find the words to reply to the compliments. This was a deserved success for his factory, and an appropriate prize for a marque that for a little while has not enjoyed an easy life but nevertheless has glorious traditions."

Perhaps the most telling comment, at least in terms of advertising merit, came from the principal daily sports journal, "La Gazzetta dello Sport": "The Benelli successes have been confirmation of the merits of an inspired motorcycle. In fact, one of the production Leoncino models as available to the public won the race." If that latter comment was not strictly accurate, it had boundless publicity value for the Pesaro factory's sales team.

In the same season, Tartarini confirmed the model's merits with a class victory in the Milano-Taranto, a success repeated the following year by Preta. However, the Leoncino will be forever associated with the Giro, for although it was never again to achieve outright victory it went on to register class successes in 1954, 1955 and 1957 thanks to Tartarini, Campanelli and Ferrari respectively.

Sadly, one week after the Giro of 1957, the Mille Miglia was beset by tragedy. The catastrophic consequences of a sports car ploughing into a bank of spectators brought a summary conclusion to the spectacle of open roads inter-city racing in Italy. The days of "The Terrible Twins", the Giro and the Milano-Taranto, were over.

The Leoncino's principal exponents, Tartarini and Campanelli, were to forge fascinating careers. Tartarini found employment with Ducati, and in 1958 undertook a trip around the world on a 125 cc Sport, whereafter he established the Italjet concern. His erstwhile colleague Campanelli raced succesfully into the mid-1960s aboard an ancient 500 cc Gilera Saturno and Motobi machines.

The Leoncino's sporting career was not restricted to the long distance epics; it performed handsomely over a range of street circuits and hill-climbs, enabling Fattori to record five victories in 1956, and indeed in that same season it registered an international success at the Nice GP in the hands of Bozzano, and also took Silvio Grassetti to his inaugural Benelli victory in a junior championship race at Modena.

So it was that, with the demise of open roads racing, the Benelli factory decided to modify a Leoncino for Italian Formula Three racing in 1958, equipping the bike with a larger tank, a racing saddle, clip-on handlebars, and an assortment of go-faster gadgetry, offering a top speed of 100 mph.

The production Leoncino was also made available in four-stroke mode, and some of these versions found their way to the race track. However, the days of the Leoncino as a racing steed were virtually numbered, for it was up against the likes of the sophisticated desmo Ducatis.

It was greatly to the credit of the youthful Grassetti that during 1958, by dint of riding that was reported to be "tigerish", he was able to challenge Franco Farne's works Ducati in a number of Italian championship rounds, and even beat him at Busto Arsizio.

It should also be recorded that the Leoncino, in virtually roadgoing trim, was a handy racing tool for privateers, being both cheap to maintain and reasonably competitive. One such rider was Fernando Bruscoli, whose career had embraced an attack on the Milano-

A 1950s 125 cc Motobi, Misano 1989.

Taranto trek in 1950 aboard a 125 cc MV Agusta, and who was typical of the local enthusiasts, competing in the popular street races that took place in the seaside resorts to the immediate north and south of Pesaro.

Like many Pesarese, motorcycling was in Bruscoli's blood. His father had known the Benelli brothers, and a young cousin was to achieve world championship status as a manufacturer in the years to come: Giancarlo Morbidelli. For the 1956 season, Bruscoli's uncle provided him with a Leoncino, and he recalled his year's sport: "To be honest by then it was outclassed as a racer by Ducati machinery as it was lacking in top speed and seriously outdated, but it was economical, reliable and lots of fun for privateers".

Motobi
Although perhaps not strictly a component part of the Benelli saga, nevertheless a review of the racing history of the literally fraternal Motobi marque is a fully justified digression.

The first-born of the Benelli brothers, Giuseppe, was an inventive designer who had taken great pride in the growth of the family's business from its early days in a tiny workshop, under the guidance of his mother, to its prominence as a major manufacturer by 1939.

He lacked the force of character to propel the company ever onwards; that role fell to his dynamic brother Filippo, upon whose financial acumen was based the firm's commercial success.

By contrast, Giuseppe was of mild-mannered, scholarly disposition and he was distraught at the devastation inflicted on his life's work by the departing German forces during the final days of their occupation of Pesaro. His designs were stolen or lost, the workshops were razed, the machinery was destroyed; for Giuseppe, over thirty years of endeavour had been wiped out. It was a body blow from which he never really recovered.

In 1947, sick of spirit, Giuseppe fell ill and was unable to assist his brothers as they strove to rebuild their business empire. In 1950, following a number of seemingly inevitable disagreements with his brothers, Giuseppe decided to start afresh, intending to establish anew a business empire with his sons Luigi and Marco. The enterprise was based in premises in Pesaro, and in its formative years produced a range of two-stroke lightweights which featured the spine frame that was to become characteristic of the marque, called "B" and subsequently Motobi.

For half a dozen years the Motobi concern inevitably was primarily pre-occupied with commercial survival, and its racing ventures were as of necessity restricted. The most notable success attained during this period was Silvano Rinaldi's class victory in the 1955 Milano-Taranto at over 60 mph, aboard a 250 cc two stroke Gran Sport Speciale.

The foundations of the company's racing fortunes had, however, been laid in the previous year when Giuseppe had proposed a four-stroke engine. In fact, much of the design work of this motor can be attributed to Piero Prampolini, who had worked on the most recent sohc Mondial road bikes.

Prampolini's first engine was a 125 cc affair, and it retained the egg-shape of its two-stroke predecessors that was to be the trademark of Motobi engines henceforth. Giuseppe was however the presiding genius of the infant concern. He was renowned for his fertile and

An immaculate 125 cc Motobi; the Cattolica paddock, 1991.

A Primo Zanzani-prepared semi-works Motobi pictured during the tour of Argentina.

Umberto Masetti, during his spell in South America, racing a 125 cc Motobi

A 125 cc race over South American streets; local rider Guillermo Carter (Morini) leads the celebrated Masetti (Motobi).

creative engineering. For instance, in 1953 he came up with a highly unusual four cylinder power plant that was in effect two vee-twin engines married onto a common crankshaft with the cylinders staggered, thereby producing a semi-radial design. The proposal was that the engine would be used to attack a number of car speed records, and could possibly be adapted for motorcycle racing thereafter. Alas, such lofty ambitions were merely pie in the sky and predictably they came to naught.

Giuseppe, savaged by the trials of the last fifteen years, died in 1957. The Motobi history was, however, far from finished; instead a new chapter began when there arrived in Pesaro an enthusiast who was to be at the heart of Motobi and Benelli exploits for the next 13 years: Primo Zanzani.

Born in 1923, he had served his engineering apprenticeship at Forli airport, working on Caproni aeroplanes during the week and racing a 125 cc Morini at the weekend. Zanzani worked in the Laverda raceshop for five seasons, during which he secured a class victory in the 1954 Giro d'Italia. At the end of 1957, Zanzani decided to retire from the tracks and found that, in the wake of Giuseppe Benelli's death, he was offered the post of Motobi's technical adviser. So it was that he took himself off to Pesaro.

Henceforth, a pushrod 175 cc bike was pressed into service in the popular production based series that was in vogue in Italy (for MSDS, Macchine Sport Derivate dalla Serie, the equivalent of the "Sport Production" machines of the early 1990s). Measuring 62 x 57 mm, the virtually horizontal cylinder relied on two valves and breathed through a single 30 mm Dell'Orto carburettor. The frame was still the idiosyncratic pressed steel spine affair of the early two-strokes. In their production racing guise these machines retained their lighting systems, number plate and kickstart. Similarly offered for sports use were the 125 cc model, with dimensions of 54 x 54 mm, and the 250 cc bike, that was simply a 175 cc machine bored out to 74 x 57 mm.

By 1960, the Motobi factory was offering its models in genuine racing trim. The frame was strengthened, being bolstered by the addition of small diameter tubing. Oldani brakes and Ceriani Grand Prix telescopic forks were fitted as standard, the tank, now available in plastic, was elongated and lowered, and the saddle shortened. Finally, an abbreviated dolphin fairing could be purchased.

Zanzani described the Motobi racing range thus: "They were not really racers. Every single bike was in essence one of the production machines but elaborated to a very high standard. My refinements were quite sophisticated and I can tell apart a works Motobi racer from a production bike that somebody else has converted to racing specification."

The marque's foremost track successes were probably garnered in South America in the late 1950s and early 1960s. The tiny factory secured a niche in the continent and indeed its biggest market was in Argentina. Hence, in 1960, the Argentinian importers arranged that Zanzani should take out a couple of bikes for the country's season, that lasted a mere two months. So although Motobi never ran an official works racing squad, Zanzani was able to experiment and develop his ideas while his products raced at Cordoba, Buenos Aires, Mendoza, Mar del Plata and Rosario, with the organisers meeting his expenses.

Thus firmly rooted in the South American racing culture, the Motobi name was kept to the fore in the continent thanks to a host of enthusiasts who campaigned the lightweight steeds in the ramshackle street races of the time, one of whom was no less a figure than Umberto Masetti who had emigrated to Latin America in the late 1950s.

A shot of a spine-framed Motobi, being a converted road-bike.

POST-WAR MISCELLANY

The Motobi raceshop in the 1960s
(Michael Dregni)

250cc Motobi, Giro d'Italia 1992

A two-stroke 50 cc Benelli in use over a South American street circuit in the early 1960s, ridden by Guillermo Carter.

The works dohc tiddler racer in the Benelli raceshop (Michael Dregni).

In Italian racing circles the 175 cc Motobi became established as a machine of some repute, capturing no less than fifteen junior category or mountain championships between 1959 and 1972.

The 125 cc Motobi, capable of 17 bhp at 11,000 rpm by 1965, appropriated another seven national titles, including a junior championship for local boy Eugenio Lazzarini in 1967. Bertarelli's lowly 18th spot in the 125 cc Italian GP of 1968 was however probably a fair reflection of the bike's status at international level.

Although it had four national titles to its credit during the late 1960s, the 250 cc machine was probably the least succcessful of the bunch. As a humble road bike in origin, tuned and revised on a shoe string budget, it could hardly hope to compete against the likes of the exotic Honda six or the Yamaha four cylinder machines that were virtually monopolising the GP racetracks.

Motobi returned to the Benelli fold in 1962 when Luigi and Marco Benelli came to an accord with their cousins but nevertheless production of Motobi motorcycles as a distinct marque continued, and indeed Zanzani remained in charge of the development of the production racers.

At just about the time of the merger of the marques, both were prominent on the track. So it was that when Provini won the Spanish GP at Montjuich Park in 1963 astride the 250 cc Benelli, the team's racing manager Sanchioni claimed: "The Benelli racers are unbeatable on all fronts, both in Italy and abroad because we have bombs." Flowing from this remark was the decision to paint the Motobi racers in a "bomb" colour; the precise tone was matched with figures on the War monument "Ai Caduti (to the Fallen) di Pesaro" that could be found in Piazza Mosca.

For some years, although there was no official racing participation, works bikes were in effect provided to a number of private teams. Prominent amongst these teams was the Scuderia (stable) Imperiali, run by the Roman enthusiast Augusto Imperiali, which captured three national titles for Motobi.

In 1969 a halt was called to Motobi's racing endeavours. Primo Zanzani immediately spotted a niche, and duly left the company to become self-employed. He prepared and marketed kits to convert road-going Motobi bikes to racers, offering pistons, valves, pushrods, tanks and even completely new frames. Thus furnished, Motobi racers remained modestly competitive into the early 1970s.

Although not blessed by the factory hierarchy, mention must be made of an eccentric special, produced by the Swiss Motobi enthusiast Werner Maltry. In 1962 Maltry had won a Swiss title with a 175 cc engine bored out to 245 cc. For the following season his offering was a 490 cc racer, with horizontal twin cylinders, measuring 74 x 57 mm. He had mated two Motobi engines and provided extensive finning to produce a smooth outline. Primary drive was taken from between the cylinders, there was an in-unit six speed gearbox, twin 28 mm Dell'Orto carburettors were employed and ignition was by battery and coil. The cycle parts were typical Motobi: a tubular spine frame, Ceriani telescopics at the front and a conventional swinging arm at the rear, with Oldani brakes.

Maltry claimed 50 bhp at 9,500 rpm and a dry weight of 230 lb. for his creation, which the press greeted with a degree of scepticism. Alas, his proposals to produce half a dozen racers for sale to privateers were but pipe-dreams. During 1965, the Maltry special appeared in practice for a handful of national races, entrusted to Paolo Campanelli, but it was never raced in anger and thus disappeared into oblivion.

50 cc Racers

Such was the growth of 50cc tiddler racing that for 1961 the FIM instituted the Coupe d'Europe, which was dominated by Hans Georg Anscheidt courtesy of a factory Kreidler. The success of the series guaranteed world championship status for the category the following year.

Although not works supported, a few two-stroke Benellis were to grace the world series in 1962.

POST-WAR MISCELLANY

The 60 cc racer.

The Pesaro factory's 49 cc Sport model had already been campaigned by privateers for a couple of years, with the Belgian Schaatsbergen winning his national title in 1960 aboard a dolphin-faired version. The usual Benelli racers were essentially little more than standard mopeds souped up for track use but in 1961 the factory decided to produce a small run of specials, although in fact the bikes that emerged were little out of the ordinary.

About a dozen frames were built for these genuine racing motorcycles; they were flimsy in the extreme and featured a solitary tube forming the machine's spine, with the engine slung beneath it. The forks and wheels were poached direct from the production moped, with tyres, front and rear, of 18 x 2 in. and drum brakes of miniscule proportions.

Similarly, the moped's two-stroke engine was employed and the solitary obvious refinement was that the pedals were removed, although additionally cooling slots were cut into the magneto covers. The engine measured 40 x 39 mm, was inclined forwards at about 45 degrees and ran through a four-speed gearbox. The gear change was operated by a clumsy cable-operated twist-grip on the clip-on handlebars.

Power was of the order of 7 bhp at 9,700 rpm. Unfaired the bike was good for about 68 mph but, when clothed with a neat dolphin fairing, in red and silver in genuine fire-engine mode, it was capable of 75 mph, slightly in excess of the best that an Itom could muster but well down on the unheard-of performance levels of the little jewels from Suzuki and Honda.

One bike found its way into the hands of Fron Purslow at about the same time as he acquired the quarter-litre single-cylinder bike that Hailwood rode in the 1962 TT. Purslow's model was the only one of three Benelli entries that actually took to the starting grid of the inaugural 50 cc TT. Purslow entrusted the baby racer to debutant Ralph Bryans, who rode the bike to 15th place, covering two laps of the daunting Mountain course at a shade under 60 mph. Ernst Degner's winning average of 75.12 mph aboard his works Suzuki put the Benelli's performance into perspective.

After the TT foray Purslow's machine was campaigned in the UK by the Tohatsu importer Jim Pink but they concluded that the bike was uncompetitive and so it soon passed on to new owners. For a couple of years, two-stroke Benellis could occasionally be seen at the rear of the pack in the classic races. For instance, one finished in 15th spot in the Dutch TT of 1962 and in the following year the Purslow model was ridden, and retired, by Alan Hutchings in the TT.

Although they never graced the international stage, Benelli did actually build two four-stroke tiddlers, one of 62 cc and the other a genuine 50 cc version. The first, with its upright cylinder and 44 x 40.5 mm dimensions, was merely one cylinder off the new 250 cc four-cylinder racer, with the gear train to the dohc assembly running up the right hand side.

The 62 cc machine was really nothing more than part of the development programme for the quarter-litre engine, enabling Savelli to try out his theories on a mobile test bed. By contrast the 50 cc version was plainly intended as a serious racing tool and Silvio Grassetti gave it a couple of outings in local races during 1963. The motor was, in typical Italian style, festooned with external oil pipes and had a well finned sump. It relied on battery and coil ignition. The cycle parts were conventional with a simple duplex cradle frame. However, Savelli's departure in 1963 left the design team heavily overburdened and Grassetti's dismissal at the end of the year effectively sounded the death knell for the tiny racer. It was refurbished towards the end of the decade for use in hill-climbs, and was provided with a slim and elongated tank and an elegant dolphin fairing, both finished in the dark green colour that was then in vogue for the team's racers.

Chapter Six: The 250 cc Four: 1962 - 1969

The much heralded quarter-litre challenger from Pesaro underwent an uncommonly lengthy gestation period. The detailed design work was principally that of Aulo Savelli, who started work on the project during 1959 and he unveiled the prototype to the press in June of the following year.

At the heart of the beast were the aircooled four cylinders, each measuring 44 x 40.6 mm for a capacity of 246.8 cc. The two valves per cylinder, with coil springs, were activated by a dohc assembly and drive was by a gear train between the middle two cylinders. Savelli departed from traditional Gilera/MV Agusta thinking in that the four cast iron cylinders were vertical, instead of being slightly inclined. Primary drive was by gear.

Giovanni Benelli claimed 40 bhp at 12,000 rpm for the engine that had a compression ratio of 10.5:1. It breathed through four 20 mm Dell'Orto carburettors, while ignition was by battery and four coils. Also featured were a dry clutch and an integral six speed gearbox. Dry weight was of the order of 270 lb.

The frame was a duplex tubular affair but its twin top tubes were paired just behind the cylinder head and then ran forward as one to the steering head. The tyres were 2.50 x 18 in. at the front and 2.75 x 18 in. at the rear. The petrol tank was short and squat, while an inadequate looking saddle seemed to have been perched as an afterthought at the extremity of the frame.

There was an oil tank slotted into the gap between the frame's rear downtubes, the rear wheel and the saddle; it looked as though it had been purloined from the Ted Mellors bike without further ado. So it was that, at its inaugural appearance, the Benelli multi gave the impression of being a haphazard collection of components, rather than an integral design. Perhaps this was not altogether surprising for, as "Motociclismo's" Carlo Perelli reported, the machine had been built in less than three months.

The machine was tested extensively by both Spaggaiari and Grassetti, despite which it was patently not ready for competition in time for the Italian GP at the tail end of 1961. The much awaited debut took place at the Imola Gold Cup meeting, the precursor to the 1962 season. The machine had undergone substantial revision over the twenty months since its press showing. The oil tank had been ditched in favour of a sump which was intended to solve a cooling problem and also lower the centre of gravity.

Furthermore, the chassis had been tidied up and slightly wider section tyres were now employed. An Oldani front brake had also replaced a Benelli version. As a sop to traditionalists, the fairing/tank had reverted to a bright red colour scheme, instead of the two tone green that had decorated the bike in testing to hoots of "insipid". The red machine was in fact built in 1962 and featured a Lucas magneto mounted forward of the engine instead of the coil system of the predecessors.

The Benelli's form was impressive at Imola, as Grassetti mixed it with the Honda four of Tom Phillis and Provini's Morini. Sadly, Grassetti blotted his copybook by missing a gear and bouncing a valve. The Benelli team leader did, however, render ample amends at the May Day Cesenatico event. Having come unstuck and crumpled his fairing, Grassetti set about his task with a vengeance, outspeeding Phillis and Redman to victory.

A week later, at the Spanish GP at Montjuich Park, astride the updated old green 1961 machine, Grassetti hounded the awesome Honda squad. He was in a challenging third position when the Pesaro multi was beset by the points floating open at high revs.

Grassetti competed at the prestigious summer San Remo international but crashed in trying to stay with Provini on the single cylinder Morini. A year that had started so promisingly ended ignominiously when the over-zealous rider blew up both the team's multis in practice for the home GP, thereby non-starting.

In truth, 1963 witnessed no discernible improvement in form or fortune. Grassetti's appearances in the classics were but intermittent. He led the West German GP at Hockenheim until fuel starvation set in and so he completed the race by blowing down the petrol tank breather pipe. A fourth place finish was respectable in the circumstances.

That the problem was reliability rather than speed was confirmed at the ultra-fast Spa circuit. Once again, the Benelli had the legs of the opposition but magneto trouble put an end to the challenge.

In a mid-season effort to cure the irritating ailments that beset the machine, a Lucas transistorised ignition system was fitted and a modified streamlining employed to direct cooling air onto the engine, as overheating was leading to power loss. Alas, Grassetti and Derek Minter, who was handed a Pesaro multi on a one-off basis, were outpaced in the Italian GP and were early retirements.

Benelli had reached a watershed; a promising design was in danger of going nowhere. As it was, a change of personnel occurred at this stage. First of all, the fiery Grassetti packed his bags. Born in Montecchio, near Pesaro, in 1937, he had thus far been crucial to the project but he had probably failed to fulfil his early promise. His future career would embrace spells with Bianchi, Morini, Gilera and indeed a return to Benelli's portals.

Another to leave the factory's employ in 1963 was Savelli, who set up a consultancy. Born in Castiglione di Cervia, just to the north of Cesenatico, he had obtained a degree in mechanical engineering at the ancient University of Padua. He had always been fascinated by performance engines and had converted a friend's sohc 125 cc MV Agusta to twin cam operation. This work formed the basis of his university thesis.

At Benelli, he enjoyed the assistance of quality mechanics, such as Ambrosini's right hand man Maroccini, Filippucci and local boy Ennio Lazzarini, brother of the future world champion Eugenio. Another of the team was Eraldo Ferraci who emigrated to the USA in 1967. Once in the Land of the Free, he was particularly associated with trick Ducatis and achieved celebrity in 1991 when, aboard a

THE 250cc FOUR: 1962-1969   63

The four-cylinder power plant in its first incarnation, 1960 *(Mick Woollett)*.

The 1962 model *(Sauro Rossi)*.

Team Fast by Ferraci Ducati, the Texan Doug Polen virtually monopolised the Superbike World Championship.

But perhaps most influential was Renato Armaroli who was primarily responsible for the development of the four-cylinder engine. Armaroli's previous experience was garnered at Mondial and Ducati and for a short spell he was effectively Benelli's chief engineer. However, in 1964 he left to work for the Spanish Mototrans concern, at which he commissioned Savelli to design a 250 cc four-cylinder racing engine. When the motor was built, it bore a striking resmblance to the Benelli.

Meanwhile, Savelli's business expanded and flourished. He offered occasional advice to Benelli and also co-operated in 1970 with Silvio Grassetti to produce a twin-cylinder disc valve two-stroke that unfortunately never raced.

The Benelli brothers, however, had an ace in their pack: Tarquinio Provini. Born near Piacenza in 1933, Provini had started to race using his uncle's name and licence, as he was under-age. He hit the headlines with victory in the 1954 Moto Giro d'Italia astride a 175 cc Mondial, and he rose to the summit of a world title in 1957, courtesy of a 125 cc machine from the same stable. For MV Agusta, he collected the 250 cc world championship in 1958. For 1960 he signed for Moto Morini, to develop their 250 cc racer, and three years later the so-called "world's fastest single" was pipped to the title by a mere two points by Honda's Redman. However, Provini was in dispute with Morini's engineers and so he switched horses to Benelli.

For the start of the 1964 season, he prompted a lighter, smaller machine. The frame was reduced in scale, and a longer thinner tank was fashioned to enable the rider to tuck himself neatly beneath the screen. Revised camshafts narrowed the power band but raised the rev limit, and 48 bhp was now available at 14,500 rpm. A seventh speed was added to the gearbox to cope with the narrow power band. Weight was 264 lb and top speed an estimated 140 mph.

Provini's initial forays with the multi were distinctly lacklustre as he suffered the humiliation of being slower than his old Morini, now entrusted to Agostini and Grassetti, at both Modena and Imola.

The European curtain-raiser to the classic series was the Spanish GP at the twisty, tortuous Montjuich Park circuit at which no less than seventeen works machines took to the grid, including the Suzuki square four, the super fast Yamaha twins and the Honda fours. But Montjuich was a circuit at which Provini excelled, with his unique style of riding. In the words of Tommy Robb: "He cranked the bike over at incredible angles and had a lurid style of racing where he would streak into a corner, then suddenly bang on the brakes. It was hair raising."

The Benelli takes the chequered flag, Montjuich Park 1964 (*Mick Woollett*).

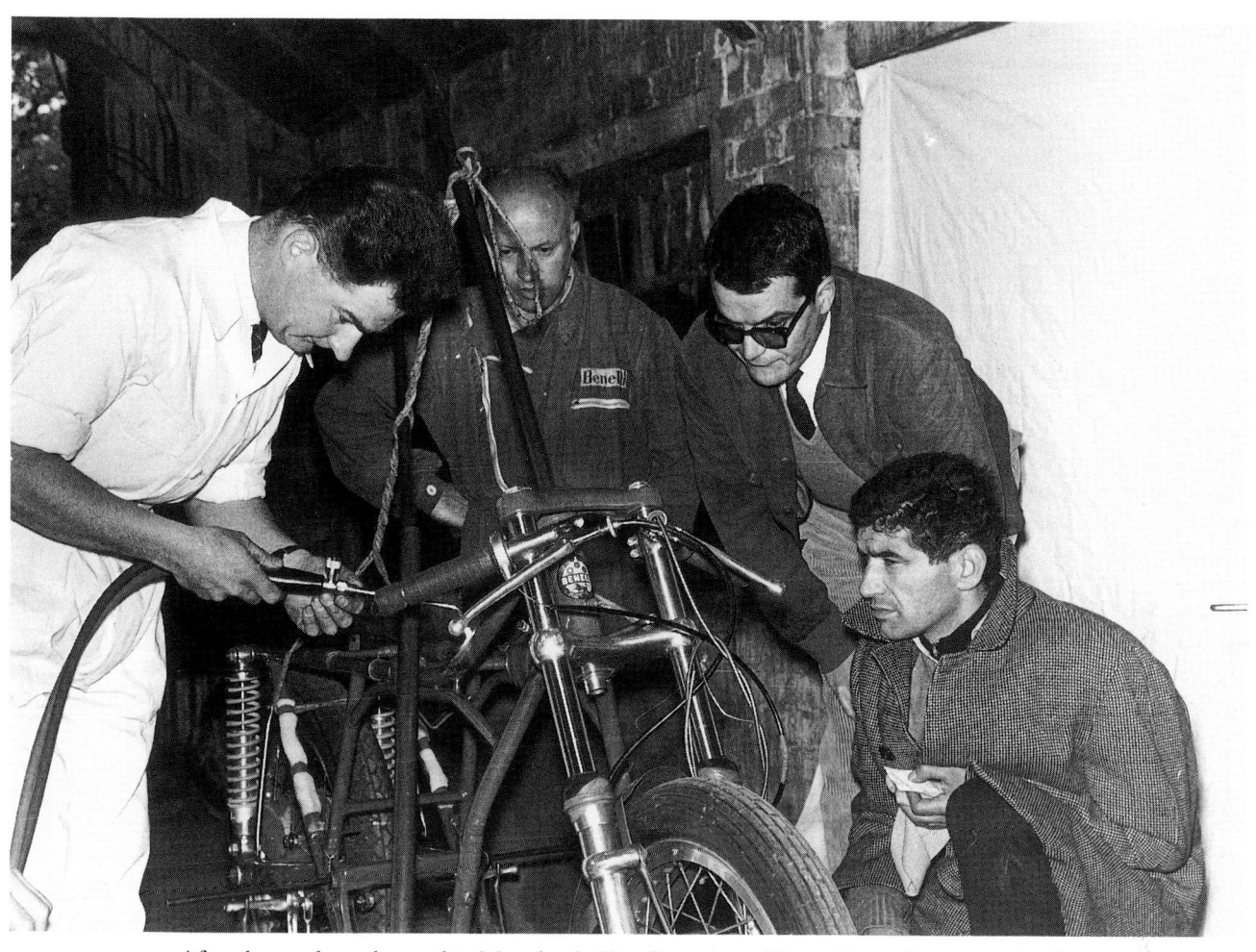

After the crash on the road to Manxland, Ken Sprayson of Reynolds, a factory mechanic, Innocenzo Nardi Dei and Provini survey the bent frame *(Nick Nicholls)*.

The 1965 model, complete with disc brakes *(Mick Woollett)*

Phil Read once described the Italian in similar terms: "Extrovert, explosive, full of life. Even on a circuit like Montjuich Park he seemed to go as fast around the bends as he did on the straights, and that's the way he lived."

Redman led for much of the race but "Old Elbows" was on his tail playing a waiting game. Sure enough, tiring of the tactic, Provini pulled out of the Honda's slipstream and headed for the chequered flag, waving to the crowds with the race in his pocket.

Unfortunately, the Barcelona victory was the highlight of an otherwise disappointing campaign. Provini failed to win any of the Italian races and Agostini rubbed salt into the Benelli wounds by regaining the title for Morini.

The classics were similarly unproductive. Indeed, the TT attack was blighted before the meeting got underway when the Benelli van on its way to the Liverpool boat crashed into a tree near Shrewsbury. Of the two multis on board, one was severely pranged, and unluckily this was the latest machine featuring a new engine with stiffer crankcases that was good for 45 bhp at 15,000 rpm.

With Savelli's departure Zanzani had been moved sideways from the Motobi race shop to oversee the Benelli stable but the ex-Mondial engineer Omer Melotti had also been recruited recently to help fill the void. Melotti was quite badly knocked about by the crash and thus Benelli's return to Manxland got off to an inauspicious start.

Fortunately, the second string multi that had been brought along as a reserve, together with an aged single, was relatively unscathed while Provini escaped the prang as he had opted to travel to the Island in his Mercedes sports car. With the help of Ken Sprayson of Reynolds, the mangled frames were repaired.

Timed over a speed trap at 140 mph, the Benelli was holding a respectable fourth place when an oil leakage betrayed the attempt at Union Mills on the sixth and final tour.

Thereafter, the Benelli evidenced reliability rather than speed with a fourth and two fifth places at Assen, Spa and Solitude rspectively. In readiness for the all-important Italian GP the frame was redesigned, emerging in Featherbed style, with duplex top and bottom tubes. A slimmer and longer tank was fitted, as was a more elegant fairing. The new streamlining was finished in a slate grey fairing, like its Motobi sister machines, with the blessing of a P.R. firm that had advised that this colour would convey the maximum impact on the race-going public.

However, when the Benelli blew up on the first lap of the GP Provini must have wondered exactly what he had let himself in for. So it was that, over the winter of 1964/65, he put in hand some more substantial modifications.

There were detailed refinements to the combustion chambers, piston crowns and camshafts, with usable power coming in at 8,500 rpm and lasting until 14,500 rpm. An eighth speed was on offer, 24 mm carburettors were introduced, while the Lucas magneto was ditched.

The Lucas unit had originally been designed for use on the lower revving Gilera and MV multis and became unreliable beyond 12,000 rpm. Despite experimentation with a form of electronic ignition, a US-made Mercury magneto provided the solution. The Mercury product began life on a four-cylinder two-stroke marine racing engine but it was successfully adapted to four-stroke use.

At the beginning of the season the racer was equipped with American-made Airheart disc brakes. However, the two small 7 in. discs had originally been intended for kart racing and were to prove inadequate to stop the 250 lb racer and hence, by the end of the year, the Benelli had reverted to Ceriani drum brakes.

In Italy Provini reaped all the available publicity with victories at Modena, Riccione, Milano Marittima, Imola, Cesenatico, Pergusa and Vallelunga, to romp home to the domestic title.

Perhaps wisely, the Benelli management steered clear of the international area for the most part, with some notable exceptions. Naturally, Montjuich Park was high on Provini's list of priorities and, true to form, after playing a waiting game, he slipped past Read's Yamaha into the lead of the Spanish GP whereupon his mechanics visibly relaxed, believing the race to be in the bag. They had reckoned without the tenacity of the English world champion, riding at the top of his form. Read had the nerve to outride the so-called Master of Montjuich and regain the lead whereupon, in his anxiety to uphold his reputation, Provini overdid it and cast himself off with only a mile to go the flag. A predictable aftermath of a display of Latin temperament and recriminations ensued.

Provini also ventured to Manxland, for what was then the most important race series on the programme. The celebrated race newspaper, the "TT Special", set up a speed trap over the stretch from the start down to Quarter Bridge. Both Redman on Honda's six and tiny Bill Ivy on a Yamaha twin were recorded at 129.52 mph, leaving the Benelli far behind at a relatively modest 122.48 mph. In the race itself the Pesaro bike lasted the distance to take fourth place.

Mindful of the expense, and conscious of the machine's limitations, Paolo Benelli, one of Tonino's sons who was in charge of the race shop, kept Provini out of the majority of the classics but he was able to win the Italian GP at the end of the year. The race took place in teeming rain which disguised the Benelli's lack of horses, and both Read and Duff were betrayed by their Yamahas. A second Benelli was entrusted to the veteran, tiny Remo Venturi. In fact, Venturi was a faller on the first lap but remounted to make up lost ground and take third spot behind Provini and MZ's Rosner.

Over the winter, a three valve head was produced, with two inlet and one exhaust valves. Although this offered better bottom end performance there was no appreciable improvement in power and so by the spring a four valve motor was built, which took power to 55 bhp at 16,000 rpm.

Once again, the domestic programme was an unqualified success, as Provini rode more or less unchallenged to six victories and another title. His appearances on the international stage were, as before, spasmodic. At his favourite hunting ground of Montjuich, he ran second behind Honda's new recruit Mike Hailwood until a couple of cylinders cut out.

Sadly, the 1966 TT, delayed until August by the seaman's strike, witnessed the untimely end to Provini's legendary career. On the

Provini rounds Ramsey's famous hairpin, TT 1964 *(Nick Nicholls)*

Provini on the grid, West German GP, 1964 *(Mick Woollett)*

Thursday morning of practice week, Provini was moving at some 140 mph near Ballaugh when he was inexplicably projected out of the saddle like a missile.

The cause of the accident was never made public. A suggestion was that Provini was caught out as he emerged into a patch of bright sunlight. However, years later Provini proferred another, more bizarre explanation. On the Wednesday evening practice stint, the engine cried enough. Instead of rebuilding it, Provini and his mechanic knowingly botched a repair and trusted to luck. Next day, the engine exploded, a chunk flew into the gearbox and locked the rear wheel.

Whatever the origin of the accident, its consequences were disastrous. Provini's back was broken and it was feared that he was paralysed. Fortunately, after months in hospital, virtually encased in plaster, he recovered but his racing days were gone. Thereafter, Provini devoted himself to the development of his business, Protar, which made scale model replicas of GP machinery.

The team's supremo Paolo Benelli and the racing manager Innocenzo Nardi Dei embarked on a hasty search for a top-class replacement. No less than three of the quarter-litre models were entered for the Vallelunga round that closed the Italian championship season in October 1966 and indeed Ballestrieri rode one to victory with Bergamonti taking another to second place.

It was, however, the third applicant, Renzo Pasolini who, with his victory aboard a 350 cc model, impressed sufficiently to be signed up for the 1967 season. In fact, Pasolini was rarely entrusted with the 250 cc machine during that, his first, season with the factory, although he did collect one domestic title round at Milano Marittima. His principal task was to develop the Junior bike, because Paolo Benelli realised full well that the 250 cc version was outclassed by the Japanese steeds.

So it was that the Lightweight category Benelli was restricted to outings in the Italian championship in the hands of an old favourite, Silvio Grassetti, who had returned to the fold. As chance would have it, his greatest adversary throughout the season was to be the recently rejected Angelo Bergamonti on the ex-Agostini twin cam Morini.

At Vallelunga the chequered flag was being unfurled in readiness for Bergamonti when the Morini's battery failed and the engine was silenced. He coasted to the line but this allowed Grassetti to take a victory that he could not have hoped for. The second title round was held at Zingonia, near Treviglio. Pasolini and Grassetti were constrained to retire, leaving their relatively unknown opponent to register a victory.

Hence, the title hinged on the outcome of the final race of the series, in September at Pergusa. Grasetti's comfortable victory was to be of no avail. Bergamonti rode tactically, refusing to be embroiled in a scrap with the Benelli duo, content in the knowledge that his

Provini used a 250 cc bike for the 350 cc Junior TT, 1965. Here he is pictured during practice.

Eraldo Ferracci (holding the gears) and Provini
(Michael Dregni).

runner-up spot was sufficient to clinch the Italian title.

In 1968 Pasolini was entrusted with the elderly quarter-litre machine with which he captured the domestic title. Internationally, the machine's appearances were spasmodic. At the beginning of June, Pasolini journeyed to Manxland, together with Domenico Benelli and two mechanics, principally to challenge for the Junior TT. A single 250 cc bike was taken along, primarily as a practice hack, but it was perceived that only the four-cylinder Yamaha machines of Read and Ivy were faster and they were believed to be unreliable.

So Pasolini took his machine to the starting line on the Glencrutchery Road and Read's retirement allowed the Italian ace to record a runner-up spot.

Reviewing the Benelli, the doyen of journalists Vic Willoughby concluded that the factory's engineering development was on a par with MV but, in an understanding of four valves per cylinder technology, there remained a considerable gap to Honda's achievements. Over the last few months Benelli had squeezed 46 bhp at 14,000 rpm from the engine - as against a probable 60 bhp for the 1967 Honda, and a likely 65 bhp for the Yamaha.

At 305 lb, the Benelli was penally heavy into the bargain. Hence, the results of "Motor Cycle's" speed trap at the Highlander were not altogether surprising: the Benelli clocked 134.9 mph whereas Bill Ivy's race-winner registered 147.5 mph.

The only other venture into the points was a couple of weeks later at Assen's Dutch TT where Pasolini took third spot over two minutes behind Ivy and Read.

However, at season's end Yamaha followed Honda and Suzuki into retirement, leaving the coast clear for Benelli, for henceforth the opposition at world level would consist of an under-financed collection of third category factories - such as Ossa - and impecunious privateers aboard production racers.

The Pasolini-Benelli combination was virtually unbeatable throughout 1969's domestic season, with successive victories at Rimini, Modena, Riccione, Imola, Cesenatico and Milano-Marittima. Such was Paso's form, and indeed the Benelli's reliability, that when they turned their attention to the world championship they were white-hot favourites. In the opening joust, at Jarama, a slow-starting Pasolini set a record lap as he hunted down Herrero's Ossa, only to crash into the guard rail, precipitating his retirement. Alas, Pasolini then broke a collarbone when he pranged his 350 cc bike in practice at Hockenheim and Benelli's tilt at the crown was in jeopardy almost before it began.

Walter Villa and Eugenio Lazzarini were called up for the third round at Le Mans but both were retirements from the Lightweight fray. Trying to salvage something from the wreckage, Domenico Benelli contacted Read and offered him machines for the prestigious TT. The Luton-born champion duly took over Paso's entry.

Once in the Island, Benelli also recruited Kel Carruthers. The Australian, who had risen to prominence with third place in the 1968 350 cc world championship, telephoned Aermacchi's headquarters and secured his release from his TT ride.

As arguably the team's better chance, Read was given the latest bike, the four-valve version, that was by now producing 50 bhp at 15,000 rpm. It was however acknowledged that the older machine ridden by Carruthers, with the two-valves per cylinder head, was more manageable, even if underpowered.

Right from the start of the Lightweight TT, Carruthers, with clear roads ahead of him thanks to having drawn racing number one, shot off into a lead that he was never to relinquish. He won what was to be his only Tourist Trophy at a speed of 95.95 mph with a fastest lap at 99.01 mph. Carruthers came home some 3 minutes ahead of Frank Perris on a Crooks Suzuki.

Read's engine expired in the closing stages. The English world champion did not mince his words and candidly described his bike as "useless" which probably did not endear him to the Benelli staff. In fairness, he was probably amply justified. "Motor Cycle's" speed trap recorded Woodman's MZ as quickest through at 138.5 mph and Read's Italian machine a lowly seventh at 129.5 mph.

The predictable upshot of the Manx escapade was that Read's sojourn in the squad came to an abrupt end and Carruthers was signed up for the remainder of the title races, on the basis that he could continue riding for the Aermacchi team in the 350 cc and 500 cc classes.

Carruthers did not immediately step into the team's spotlight because Renzo Pasolini returned to take up the cudgels as the number one rider at the Dutch TT at Assen. Although Gould started like the proverbial rocket his Yamaha soon began to misfire, enabling Pasolini to register a welcome victory with his new team-mate making it a one-two for the Benelli concern.

The squad suffered a setback at the ultra-rapid Spa circuit in the next classic. Pasolini's machine expired on the notorious Masta straight, suffering what the press dubbed a "Masty death." Carruthers challenged for the lead but his engine lost its edge, leaving him in third spot behind the season's revelation, Santiago Herrero aboard the surprise package, the single-cylinder Ossa, and Rod Gould.

The East German round was tragically marred by the death in practice of the popular Bill Ivy, who was probably thrown when his two-stroke 350 cc Jawa seized. In a damp race, Pasolini played a waiting game, for once restraining his attacking instincts, to sit behind

THE 250cc FOUR: 1962-1969

Derek Minter, 1966 250 cc TT, during practice.

The quarter-litre model on semi-permanent loan to Provini. The TT Lap of Honour, 1985.

Herrero until the final mile of the Sachsenring when he pulled out of the Ossa's slipstream and took the chequered flag by a veritable whisker. Carruthers, once again a victim of an off-song motor, limped home to fifth spot. The championship stakes were hotting up; the table stood at Herrero 77 points, Kent Andersson (Yamaha) 57, Carruthers 43 and Pasolini 30.

The results of the next round, held over the closed roads at Brno, set the title chase ablaze. The Benelli team leader rode through to his third victory in four races. The Benellis were now closing on both the lone Ossa and the Swedish privateer Andersson who had not won a round since the second race at Hockenheim.

Alas, the team's confidence was shattered at the beginning of August at Imatra. The 250 cc race was an absolute scorcher. Rosner aboard his MZ, got away to a flyer and dragged the field round until he crashed. Then, Herrero managed to slide off at a slow corner. A couple of corners on, Pasolini crashed heavily at a fast right hand bend, sustaining injuries that were to put him out of action for the remainder of the season. So it was that Andersson romped to victory, with Kel Carruthers languishing in fourth place, plagued by an underpowered motor. The Australian, although more than 20 points behind Herrero, was henceforth to enjoy his choice of machinery in an attempt to retrieve the deficit.

Benelli's immediate concern was with the Italian title, for Bergamonti, riding for Aermacchi, was in a position to deprive the absent Pasolini of the spoils with a good result at San Remo. Hence no less than three of the Pesaro mulits were entered for the final race of the domestic series. Roberto Gallina and Lazzarini could not hold Bergamonti but the Aermacchi rider's third spot, behind Herrero and Walter Villa on the leading Benelli, was not quite sufficient. He finished the season equal on points with Pasolini but the latter took the title thanks to his victories.

The Benelli team, accompanied by Mimo, still a race enthusiast, then headed for Dundrod. As Carruthers attempted to impose his authority on the race, Herrero, desperate to stay with his Benelli rival, overdid it yet again and pitched himself through a hedge at Wheeler's Corner. The Spaniard must have rued his over-exuberance, as Carruthers and Andersson finished one-two to eat into his championship lead.

The penultimate title encounter was held almost in Benelli's back yard, at Imola's Dino Ferrari circuit. It was the scene of another Phil Read-inspired escapade. Noting the Englishman's speed in practice on his private Yamaha twin, Benelli wisely sought an insurance policy by offering him a ride, provided he dutifully kept in station behind Carruthers. Naively, no money was on the table and, inevitably, Read, who was first and foremost a professional racer, curtly declined the offer.

Indeed, vexed by the Italians' attitude, the impish Read told them that a four-cylinder Yamaha was up his sleeve and, to their consternation, he duly added a couple of dummy pipes to his machine!

In the race, Read realised that the Benelli was the fastest machine on the track but the Yamaha's superior acceleration enabled him to indulge in a battle of wits with Carruthers and pip him to the line by inches. Gilberto Parlotti, who was astride the back-up Benelli, dropped his bike when in third spot.

After eleven rounds, the three title contenders ventured to Opatija for the decider neck and neck: Herrero, with 83 points, was just one ahead of both Carruthers and Andersson. The Opatija circuit was regarded as a downmarket Monte Carlo. The course weaved along a coast highway and was connected, via two acute hairpin bends, to a cliff-top road. Both Carruthers and Parlotti made clean starts but they were soon joined by Herrero who was riding with his left wrist encased in plaster, a legacy of his spill at Dundrod. After a couple of hairy moments, the tiny Spaniard finally overdid it on a fast downhill bend, putting himself out of contention, and frightening the Benelli duo into the bargain. Carruthers subsequently reported: "He came off just in front of me; scared me silly."

Phil Read pushes off at the start of the 1969 TT *(Nick Nicholls)*.

Kel Carruthers at scrutineering for the Lightweight TT, 1969 *(Mick Woollett)*.

Read in Parliament Square, Ramsey, 1969
(Sauro Rossi).

Certainly the leaders appeared to have been de-tuned by the episode, for they slowed, allowing Kent Andersson to catch them, and as the race neared its conclusion the Swede seemingly had the initiative. However, with merely a couple of laps to the chequered flag, Andersson suffered a mighty 'moment', which relegated him to third place. Although he was but a handful of seconds behind the Benellis, he could do nothing about the deficit. Kel Carruthers, with only his team-mate for company, crossed the finishing line and, at long last, took the world championship that had seemed to be on the cards for the four-cylinder Benelli from the days in 1962 when Silvio Grassetti had served notice of its promise.

In its moment of triumph the 250 cc Benelli was already on the edge of extinction thanks to the FIM's edict that as from 1970 the 250 cc class racers would be limited to a maximum of two cylinders. The Benelli had literally proved itself with its last breath.

The FIM's policy, that was extended into the other categories, was intended to encourage under-funded factories to enter the fray. Sadly, it had the adverse effect of killing off numerous designs of exceptional technical interest, such as the four-cylinder Yamahas. One such victim of this policy was Benelli's proposed 250 cc vee-eight.

Rumours of this fabulous engine began to filter throught to the press in 1965 and after the Italian GP of that year Provini confirmed that the project was afoot. He reported that the motor would be set across the frame, with one bank of cylinders lying horizontally and the other bank almost vertically. When Mike Duff visited the race-shop in April 1967 he was shown the drawings of the vee-eight and told that it was hoped to have a bike ready for the Italian GP.

Those hopes proved to be over-optimistic but Innocenzo Nardi Dei was able to reveal some technical details by the time of the TT in June 1968. He claimed that the engine would go to 20,000 rpm producing 66 bhp, and that it would run through a nine-speed gearbox. The prototype engine had coil ignition but the mammoth Philips organisation had been asked to co-operate with the development of a transistorised system. There were two valves per cylinder but it was Nardi Dei's intention that a four-valve version should be introduced shortly.

It was also announced that, thanks to the use of the very latest lightweight materials, the engine weighed only 90 lb, about half of the weight of the four-cylinder power plant, and would be 3 inches narrower into the bargain. The complete machine was expected to weigh somewhere in the region of 220 lb. On paper, at least, like so many machines, it was a winner!

There had also been mutterings that the design could have been transplanted to the 350 cc and 500 cc arenas and indeed when Phil Read visited Pesaro early in 1969 he saw what he understood to be castings for a 350 cc motor. Alas, it was never to be, for when the FIM announced its proposals the vee-eight project was still-born.

THE 250cc FOUR: 1962-1969 75

Renzo Pasolini and Kel Carruthers battle it out at the Dutch TT;
they finished first and second respectively *(Mick Woollett)*.

The start of the Finnish GP with the championship contenders in place: Herrero/Ossa (1),
Andersson/ Yamaha (2), Pasolini (3) and Carruthers (4) *(Mick Woollett)*.

Carruthers wins the Yugoslav GP to take the championship in style, with Parlotti in his wheel tracks *(Mick Woollett)*.

An ex-Carruthers 250 cc model on display at Misano, 1990.

Chapter Seven: 350 cc Exploits: 1965 - 1973

When Giovanni Benelli led his troops back into the fray in 1959, he contemplated using Savelli's new dohc 250 cc single-cylinder machine as the basis of a tilt at the Gallarate company's 350 cc crown. Benelli's initial ploy was simply to overbore an engine to take it to what was variously reported as 251 cc or 267 cc. Silvio Grassetti rode the bike on a handful of occasions albeit without conspicuous success, although he collected fifth place in the 1961 Italian GP. Giovanni's plans to introduce a fully-fledged 350 cc motor, complete with desmo head, were shelved in 1960 when Savelli's four-cylinder project was taking shape.

It was actually Savelli's multi that was to take Benelli into the 350 cc arena, starting with the Junior TT in 1965. Provini entered purely to enjoy the boon of extra practice, or at least that was the party line, which also propogated the theory that his motor was a 251 cc affair. In fact, the Benelli was an early retirement in the Manx race but by the time the all-important Monza classic came around a 320 cc engine was available, with bore and stroke dimensions of 50 x 40.6 mm. With 53 bhp on offer at 14,700 rpm, Provini rode the bike to third spot in its debut at the Italian GP.

For 1966, the same machine took Provini to a victory at Cesenatico and the runner-up spot at the prestigious Imola meeting, behind Ago's MV. Irked as usual by the MV's all-too-evident superiority, the Benelli team lodged a formal protest, claimimg that the Gallarate triple was actually a 500 cc version. Needless to relate, the objection was without foundation.

In the first of the season's classics, at the uninspiring Hockenheimring, Provini had to give second best to Hailwood's all-conquering Honda six, to register the marque's solitary points in the table. With Provini's near-fatal tumble in the TT, henceforth the 350 cc Benelli was to be entrusted to one of the legendary names of Italian motorcycling: Renzo Pasolini.

Massimo Pasolini had been a highly respected pre-War racer whose greatest claim to fame was, perversely, probably his achievements in 1956 when, aboard streamlined cigar-shaped Aermacchi-powered record-breakers, he annexed the flying mile and flying kilometre 50 cc and 75 cc marks on the Milan-Varese autostrada.

His son Renzo came to the fore in the mid-1960s when he took the factory development 250 cc and 350 cc Aermacchis to a string of rostrum positions at world level. In time his name was to become almost as inextricably linked with the Benelli marque as had been Ambrosini's nearly two decades earlier. However, the stark truth was that his was a disappointing and lacklustre 1967 season.

Benelli mechanics at work in the garages at the Douglas Bay Hotel, 1966 TT series.

Grassetti ploughs into the straw bales, Cesenatico 1967 *(Gianni Perrone)*

Paso sweeps his mount into Creg-Ny-Baa, Junior TT 1967 *(Nick Nicholls)*

350 cc EXPLOITS: 1965 - 1973

Grassetti aboard the 350 cc Benelli at Rimini in 1968. He leads Bruno Spaggiari (Ducati, 54) and Agostini (MV, 2) *(Mick Woollett)*.

Nardi Dei, Mimo Benelli and Paso with the 350 cc machine, 1968 TT *(Nick Nicholls)*.

Encouraged by Provini's showing on the 320 cc bike, the engine had been revised, with new dimensions of 51 x 42 mm, for a displacement of 344 cc. Power was now up to 64 bhp at 14,500 rpm.

In the Imola Gold Cup meeting Paso ran to second spot behind Ago. He was also to record victories at Cesenatico and Rimini. It was however his team-mate, the recalled Silvio Grassetti, who was to notch the Italian title that was re-instated that year.

On the world championship front, Paso took two respectable third places at Hockenheim and Assen. His exploits at the latter race consolidated his popularity with the fans. After Hailwood had performed his customary disappearing trick, Ago had to contend with an inspired Benelli rider, employing to full effect his lurid and frightening style - that had originated during his days as a moto-cross rider.

In the words of "Motor Cycle": "Few would have bet more than a plateful of spaghetti on Renzo Pasolini and his Benelli living with the formidable Giacomo Agostini (MV) yet there was Paso, all arms, legs and oops-a-daisy, sticking his neck out so far that Ago not only could not but dare not pass." Eventually, with the fastest lap of the race, Ago squeezed through to steal second spot.

Pasolini was then beset by engine failure at Brno and Monza, at which Grassetti salvaged second place for the team, well down on Ralph Bryans on his Honda six.

During the winter months, the engine, now boasting four valves per cylinder, underwent extensive bench-testing, running for hours at 14,000 rpm without mishap. The frame, that originally had been culled from the quarter-litre model, was strengthened, and Colin Lyster was called in to assist with applying disc brakes. Although Provini had toyed with discs, Pasolini had used drum brakes throughout 1967.

The team's confidence for 1968 was boosted by Pasolini's early season victories at Cesenatico and Imola, with the machines now bedecked in fairings with a light grey top and a light pastel green bottom. Alas, the domestic success proved to be another false dawn, for when the world championship got underway Agostini took his three-cylinder MV to a convincing victory in the opening round at the scenic and demanding Nurburgring.

The next stop for the two Italian contenders was the awesome Mountain circuit. Pasolini, unlike many Italians, claimed to like the classic 37-mile closed roads course but bewailed his lack of practice. Perhaps that was his undoing for, although he lapped in the race at over 103 mph, he was never in a position to challenge Agostini.

Mike Hailwood, who attended the TT in the unaccustomed role of spectator, adjudged that the Pesaro machine enjoyed the edge on acceleration over its Gallarate rival but that the MV probably handled better. Hailwood's comment that the Benelli was obviously as fast as the MV was borne out by "Motor Cycle's" speed trap at the Highlander. Fastest machine of the fortnight was, predictably, Ago's half-litre fire-engine triple at 157.9 mph. Next came his team-mate John Hartle's elderly 500 cc four-cylinder MV at 152.5 mph. Interestingly, both Ago's 350 cc model and Paso's Junior mount matched this speed.

Pasolini rounds Quarter Bridge during the 1968 Junior TT *(Nick Nicholls)*.

350 cc EXPLOITS: 1965 - 1973

Vic Willoughby was not convinced that the Benelli team had the expertise to exploit a four-valve per cylinder head. Motor cycle sport's most respected technical journalist recognised that the switch to the new heads had probably produced an extra 5 bhp and cured a valve breakage problem that had beset the bikes in 1967. On the debit side, the useful power band had been severely narrowed, so that the seven-speed gearbox was barely sufficient. In exoneration of Pasolini's efforts, Willoughby also pointed out that at 330 lb the Benelli was hopelessly overweight. By contrast MV claimed 255 lb for their contender; even allowing for some tongue-in-cheek assertions, there must have been a considerable weight discrepancy between the two bikes.

The debate came to an abrupt end at Assen when Pasolini slid off at a slow speed corner, prompting his retirement at the pits and effectively bestowing the title on Agostini. The second and third places salvaged by Paso and Grassetti in the final round at Monza were but crumbs scavenged from Agusta's table.

For 1969, Pasolini literally took matters into his own hands, preparing his engines in the factory race-shop, and seemingly with some success, as he virtually monopolised the Italian title races, registering victories at Rimini, Modena, Riccione, Imola and Milano Marittima, and generally humiliating Giacomo Agostini into the bargain.

However, Benelli's enduring propensity for self-destruction came to the fore when the world title rounds commenced. Arriving at the Jarama circuit, that was the scene of the Spanish GP for the first time, the teams found that MV and Ossa had booked exclusive use of the track. Alleging bad sportsmanship, Nardi Dei withdrew his squad, parting with the words: "If Agostini wants to practise alone, he can race alone."

When Pasolini then pranged his 350 cc bike in practice at Hockenheim, his expected challenge for the world title was effectively torpedoed and the team concentrated on the Lightweight bike for the remainder of the season.

The Nurburgring, 1970. Agostini (301) has pole position; Carruthers (311) languishes on the second row *(Mick Woollett)*.

With the 250 cc multi outlawed by the FIM's generally unpopular regulations, Benelli's hopes for 1970 were once again pinned on the 350 cc model. There were already tales that a brand new machine was being designed and Nardi Dei opened up negotiations with Agostini, offering a rumoured £23,500 for his services, and being rather upset when informed that his Agusta stipend was handsomely in excess of that!

The now redundant 250 cc bikes were re-built as spare Junior machines and Carruthers and Pasolini were retained. Unfortunately the team's preparations were devastated by a series of strikes over the winter and a serious fire in February that destroyed a store of precious metals, including a quantity of titanium that had been destined for use on the racers.

Pasolini and Agostini shared the spoils in the curtain-raisers to the 1970 season over the Adriatic circuits but, true to the script, the Benelli endeavour fell apart at the seams once the serious business began. Trying to learn the intricacies of the fearsome Nurburgring, aboard a 650 cc twin-cylinder roadster, Pasolini fell off yet again, injuring his ankle and putting himself out of two title races. Carruthers rode his Benelli to second spot in the German GP but he was far adrift of the rampant world champion.

Matters hardly improved at Opatija. Ossa's star Santi Herrero was approached to replace the absent Paso but, but with rider safety gradually coming to the fore of the collective consciousness, Nardi Dei withdrew the offer, believing that the tiny Spaniard could not reasonably be expected to get to grips with a multi on such a dangerous circuit.

Hence, the back-up Benelli was entrusted to Gilberto Parlotti whose exuberance took him into the lead of the Yugoslav GP and then, accelerating away from a tight hairpin, he dropped his bike and, having broken the gear pedal, was forced to retire. Carruthers rode prudently to another runner-up placing.

Neither Pasolini, back in the fray, nor Carruthers managed to challenge Ago at the TT and both were retirements in mid-race. The meeting was marred by the death of Herrero who had crashed his 250 cc Ossa. Both Carruthers and Nardi Dei expressed their doubts about the famous Mountain course and indeed the Benelli factory was never to return. At this point Carruthers opted for Yamaha power, leaving Pasolini to carry the Benelli standard, which he did with successive second places at Assen, Sachsenring and Brno. However, the Pasolini-Benelli relationship was at breaking point, prompted by the latter's courting of Ago and, reputedly, Alan Barnett.

When the linkage to two of the four carburettors failed, and thereby deprived Pasolini of victory at Imatra, a public and eminently predictable row erupted. A distant third at Monza, behind not only the perennial rival Agostini but also Bergamonti, was Pasolini's final fling. When he subsequently began negotiations with both Aermacchi and Jawa, it was hardly surprising that Paolo Benelli sacked him.

By 1971 the Benelli company was in dire financial straights and beset by industrial unrest. The announcement was made that the

Above: The 1970 350 cc model *(Mick Woollett).*

Left: Pasolini during practice for the 1970 TT *(Sauro Rossi).*

350 cc EXPLOITS: 1965 - 1973

Carruthers during a practice session for the Junior TT, 1970 *(Sauro Rossi)*.

The second generation 350 cc racer; Walter Villa gives it an airing. *(Sauro Rossi)*

factory would not be racing during the forthcoming season. However, in August the Tonino Benelli Club organised an international over a new road circuit on the edge of Pesaro and the Club's president, Paolo Benelli, produced one of the ageing 350 cc racers for Hailwood. The champion's return attracted a crowd of 60,000. He held the lead initially but his old adversary Agostini was merely toying with him and soon pulled away to win.

During the course of 1971, the Benelli family sought financial salvation by selling out to the Argentinian entrepreneur Alejandro de Tomaso. Sadly for enthusiasts, de Tomaso was not inclined to venture his capital in a full-scale racing assault and so it was that the career of a new 350 cc racer was to be severely curtailed.

Designed by Aurelio Bertocchi, the ultimate 350 cc Benelli - of which probably two were built - relied on a dohc motor, but instead of being upright, as had been all the multis over the preceding decade, the engine was slightly inclined forwards. It featured four valves per cylinder and the camshafts were driven by a central gear train. Other touches were the magneto, now placed behind the cylinders, a six-speed gearbox and 32 mm Dell'Orto carburettors. Power was believed to be not far short of 70 bhp.

All the engine parts were immaculately machined; by contrast, the frame, a standard duplex affair, was less handsomely finished. The frames of the final generation racers may well have been developed in a haphazard manner, on an ad hoc race-by-race basis. Ceriani suspension units were employed and the 18 in wheels were stopped by disc brakes. At the rear could be found a fully drilled 280 mm disc, while two 250 mm units were fitted to the front. Quite apart from the disc brakes, the bike was instantly recognisable by its colour scheme. The tank was all black and the fairing was a distinctive red and white design.

Benelli hankered after the Finnish sensation Jarno Saarinen for the 1972 season but, on his signing for Yamaha, Bertocchi's new bike seemed destined to become a museum piece. However, when the Benelli Club's Pesaro international came around Saarinen accepted the invitation to campaign a brace of Benellis.

In a truly exciting 350 cc race, that had the partisan locals on their toes, Saarinen and Agostini indulged in a genuine ding-dong battle, that seemed to have been resolved in the MV rider's favour when the Benelli shed a left-hand exhaust. To the delight of the crowd, Agostini's subsequent retirement allowed the Flying Finn to take the chequered flag.

Although the victory at Pesaro had been encouraging, it was but a scant pointer to the potential of the new racer, as the race had been held over a damp track. Without Saarinen's much sought-after services for 1973, as he had understandably signed for Yamaha once again, the Benellis were entrusted to Walter Villa, with a view to contesting merely the domestic events.

The 350 cc machine's season was an unmitigated disaster. Villa suffered a gearbox failure at Modena, the magneto packed in at Misano and his only decent showing was at Vallelunga where he finished 10 seconds down on Paso's Aermacchi.

Sadly, far worse was to come. Villa rode the bike to a distant fifth in the Italian GP at Monza. It was believed that his bike had shed oil on the Curva Grande; it was not cleared and may have caused the tragic incident at the start of the next race, the 250 cc event, which claimed the lives of both Saarinen and Pasolini, two of the sport's legendary characters.

The entire Italian racing fraternity was devastated by the catastrophe, which prompted Benelli's withdrawal from the tracks. The squad reappeared for the end of season Misano international when the 350 cc machine was ridden by Roberto Gallina. That he was forced to retire was both typical of the history of the 350 cc Benelli and prophetic.

Henceforth, the 350 cc racers were consigned to a corner of the Pesaro factory and, apart from a hare-brained scheme to revive them in the mid-1970s, there they would linger, dust-laden, until a decent interval had passed, whereupon they would return to life during the classic revival of the 1980s.

The factory's final 350 cc racer, often paraded during the late 1980s by John Surtees, photographed at Misano, 1989.

Chapter Eight: The 500 cc Campaign: 1966 - 1973

As with its 350 cc endeavours, the Benelli team's 500 cc exploits were dependent on two models. First, an enlargement of Provini's 250 cc racer; second, an entirely new machine, to a design by Bertocchi. The development of both was to be haphazard, intermittent and under-financed. Hence, in retrospect, it is hardly surprising that the 500 cc Benellis made but a modest contribution to the marque's roll of honour.

Pasolini's maiden victory for Benelli, in the 500 cc race at Vallelunga at the tail end of 1966, was actually achieved on a 350 cc model but a genuine Senior bike had been prepared by the time the 1967 season came around. In all major respects, it was a replica of its immediate 250/350 cc forbears; the four vertical cylinders each boasted four valves, the dohc assembly was gear-driven, the magneto sat at the front of the engine and was driven by a shaft running from the crankshaft, there was a seven-speed gearbox and a dry clutch on the left-hand-side.

The engine dimensions were, however, now square at 54 x 54 mm, for 494 cc, and 75 bhp at 12,800 rpm was claimed, somewhat extravagantly. In essence the cycle parts were those of the 350 cc racer but both the frame and brakes were slightly more robust. Despite this, weight was reputedly a reasonable 300 lb.

The machine appeared for its debut, at Modena early in 1967, clad in a new fairing in dark green with a striking yellow stripe amidships. When Agostini retired the MV, Paso carried the Benelli to a handsome victory that was to prove to be the machine's zenith. For thereafter, although Pasolini collected rostrum positions at both Riccione and Milano-Marittima, the bike was plagued by brake problems, as the drums simply could not cope with the extra power churned out by the bigger engine.

Pasolini used the season to accustom himself to the Pesaro multis and expensive international sorties were kept to a minimum. Acknowledging Pasolini's inexperience over the fearsome Mountain circuit, Benelli proposed to offer their multis to the ex-Yamaha star Mike Duff for the TT but those plans fell apart. Hence Pasolini was the team's sole representative in the Diamond Jubilee Senior TT and he held a comfortable third place ahead of the horde of British singles until the dark green Benelli expired on the fourth lap at Sulby.

A true indication of the challenge posed by the machine may be gauged by the fact that Paso was lapping at just over the ton, at about the speeds achieved by McIntyre on his Gilera a decade earlier. By contrast, supermen Hailwood and Agostini conducted their celebrated high-speed duel in the realms of laps at 108 mph.

After the Manx sortie the big Benelli was put under wraps, seemingly for good. Innocenzo Nardi Dei explained that, as the machine was basically an enlarged 250 cc affair, the engine's 80 bhp (or thereabouts!) was putting certain components, and particularly the gearbox, under an intolerable strain.

In September of 1968, now clothed in the colours of grey and green, two machines competed in the Italian GP. Benelli had hankered after the services of Mike Hailwood at Monza but he had rebuffed their offer of rides on the Lightweight and Junior models, preferring to return to MV on a one-off basis in the 350 cc and 500 cc classes. However, when after Friday's practice Count Agusta made it clear that Hailwood would be expected to toe the party line and ride astern of Ago, the Englishman refused to play ball and was on the verge of packing his bags.

At this point, Pasolini stepped into the breach, offering to surrender his 500 cc Benelli to the English star - which Hailwood eagerly accepted. In fact, by working overnight the Benelli mechanics were able to prepare a second machine, which actually enjoyed a beefed-up gearbox. After Saturday's practice Hailwood realised that he would probably have to rely on rain to stay with the MV but when it actually came during Sunday's race it probably proved to be his undoing. He stayed with the world champion for a couple of laps but then fell at the Parabolica, caught out by the damp track. His claim that Ago had cut across him, causing him to brake harshly, led to the inevitable acrimonious exchanges between the excitable Italian teams. Pasolini, on the slower of the two bikes, finished over half a minute down on the victorious Gallarate fire-engine.

Trying to tempt Hailwood into signing a contract for 1969, Nardi Dei gave him a run on the 500 at the end-of-season Riccione international but, after two laps at the front, he was betrayed by gearbox problems and toured in behind Ago.

Rejected by Mike the Bike, Nardi Dei approached the diminutive Bill Ivy but Yamaha dissuaded him from pursuing the offer. Ironically, this opened the door for his erstwhile team-mate Phil Read who raced the Pesaro standard-bearer at Vallelunga complete with a Yamaha front brake in an attempt to cure its deficiencies. Alas, it was to no avail; he still could not stay with Ago's MV although he ran to second place. Pasolini took third spot on the 350 cc model while Alberto Pagani brought the second 500 cc machine home in fourth spot. To prove that the spirit of sportsmanship was still alive, Pagani had been offered the bike when his Linto had failed in practice.

With those unspectacular escapades, the half-baked assault on the prestigious Senior category virtually fizzled out, although Paso put the Benelli cat amongst the Agusta pigeons in the Italian GP of 1970 when he harried Agostini and MV's new star Angelo Bergamonti until his bike began to lose oil.

By 1971 Benelli's participation in the sport, and indeed its continued commercial viability, was under close scrutiny. The policy decision was taken that if the team did continue on the tracks a top-flight rider must be engaged. After tentative attempts to recruit Ago fell through, and disenchanted Pasolini into the bargain so that he trooped off to Aermacchi, Hailwood was left as the only realistic alternative.

Nardi Dei reputedly offered the Englishman an annual fee of £30,000 and control of the race-shop but Hailwood had the wit to realise that, by signing, he would have put himself under an intolerable pressure to succeed, at a time when he had not raced consistently for

Pasolini pushes off in the 1967 Senior TT (Nick Nicholls).

The 500 cc Campaign: 1966 - 1973

Paso in action during the Senior TT aboard the big Benelli (*Nick Nicholls*).

Right: Paddock shot of Paso, 1969 *(Nick Nicholls)*

Below: Phil Read tests the 500 cc racer at Vallelunga, 1968
(Mick Woollett)

The 500 cc Campaign: 1966 - 1973

three seasons, whereas Giacomo Agostini was probably at his zenith aboard the thoroughly reliable MV.

With Hailwood spurning Nardi Dei's advances, a completely new 500 cc racer, that was to have been at the forefront of the team's assault on MV's castle, remained untried, languishing in the Pesaro work-shop.

The story of Bertocchi's brainchild started at the tail end of 1969 when Nardi Dei made contact with Gerhard Heukerott, a wealthy West German privateer who was a steel salesman for the giant Krupp concern and enjoyed the backing of a syndicate of enthusiastic sponsors. With their support he had acquired, via a circuitous route, one of the thoroughbreds of the race tracks: a 250 cc six-cylinder Honda. Benelli offered a deal: they would exchange, on a loan basis, a 350 cc and a 500 cc racer for the Honda. Heukerott swallowed the bait and duly campaigned his Benellis during 1970, with a fourth place aboard the 350 cc at a Zeltweg international as probably his most notable performance.

Meanwhile, back in Pesaro, Bertocchi and his cohorts were examining the internals of the mighty Honda and the lessons were no doubt applied to the new 500 cc racers that were built. The motors were akin to those of the contemporary 350 cc siblings: the four cylinders featured four valves, with the twin overhead camshafts governed by a central gear train. A centrally disposed 12 mm sparking plug was used, four 33 mm Dell'Orto carburettors were employed and a six-speed gearbox was relied upon. Three litres of oil were carried in a deep finned sump.

Ignition was still by courtesy of a Mercury magneto, duly modified to cope with high revs. Although the factory never advertised the cylinder dimensions, they were keen to boast of 100 bhp at 13,500 rpm and indeed the red line appeared at 16,000 rpm - hence the need for the modification to the magneto. (In truth, 82 bhp was a more realistic assesment.)

The frame used 28 mm tubing and there was an oval-section swinging arm. Front forks were by Ceriani as were the rear shock absorbers. Heavily drilled 280 mm disc brakes were fitted, two at the front and one at the rear. As with the new 350 cc racer, the 500 cc version was topped off with a matt black tank and a white fairing, interrupted by a red flash. Weight was approximately 320 lb, a shade up on its MV competitor, but in top speed it gave nothing away at an estimated 165 mph.

By 1972 the challenger was prepared for battle but Benelli could not raise the funds to prise Ago from MV or Saarinen from Yamaha. It was however Saarinen who gave the bike its debut and, in retrospect, its one moment of glory at the Pesaro international in August.

On the tight circuit, the Finn put in a fastest lap at 84.76 mph and romped home ahead of Ago and Alberto Pagani on their MVs and Roberto Gallina, Paton. In truth, his victory over Agostini, although immensely popular with the locals, was hollow. The world champion had been rammed on the starting grid despite which he pulled through to challenge the Benelli until a misfire set in, forcing him into the pits. Even so, he rejoined the fray and was not too far behind Saarinen at the finish.

The 1970 Italian GP at Monza: the 500 cc class competitors included Agostini (1), Bergamonti (MV, 3), Pasolini (2), Roberto Gallina (Paton, 16) and Gianni Perrone (Kawasaki, 12) *(Mick Woollett)*

Pasolini's 500 cc machine ridden by Mauro Righi, Cattolica 1991

Paso's Benelli in the paddock at Cattolica, 1991.

The 500 cc Campaign: 1966 - 1973

With Saarinen re-signing for Yamaha, the de Tomaso-owned factory was virtually at the end of the road, being rider-less and indeed rudder-less, for by now many of the old school had left the factory to take up fresh employment with Benelli Armi in Urbino, under the guidance of Paolo Benelli.

It was Walter Villa, one of the stalwarts of the Italian racing scene, who saw an opportunity and obtained a 500 cc bike for the 1973 domestic season. His best result was victory in April at the Modena international, in Ago's absence, when he nevertheless had to fight off the stern challenge of Kim Newcombe, Konig. The status quo prevailed when Agostini returned at Vallelunga so that Villa had to settle for second place.

And so to the tragic Monza GP. In practice the Benelli had demonstrated a surprising turn of speed, not far adrift of Saarinen on the phenomenal two-stroke Yamaha and the MVs of Agostini and Read, and even up on Kanaya on the second Japanese factory multi. Of course, the race was never to be, cancelled in the wake of the catastrophic 250 cc event.

It remained for two 500 cc bikes to be carted the few kilometres up the road to Misano for the October international for one last fling. Although Drapal retired his mount, Gallina at least had the consolation of taking the flag in third place behind Read on the perennially succesful MV.

And that really was the end. Three years later the organisers of the Pesaro international approached Andreas Ippolito of Venemotos with a proposal that the ex-Saarinen models should be ridden by his outrageously talented prodigy Johnny Cecotto. Fortunately for the reputation of the black, white and red racers, the half-baked proposal was rejected.

Jarno Saarinen's 500 cc Benelli after its victory at Pesaro, August 1972 *(Michael Dregni)*

Saarinen's 500 cc racer, photographed in the paddock at Misano, 1989, next to a 350 cc MV.

Walter Villa re-acquaints himself with the 500 cc Benelli at Imola, 1989.

Chapter Nine: Postscript

During the 1970s Benelli machinery participated in the production-based classes. In endurance racing, some repute attached to the French MOC Equipe which ran a futuristic device built by Motobecane, the Benelli importers, that was powered by the glorious, transverse six-cylinder 750 cc engine. The machine was characterised by curiosity value rather than race-winning credentials.

Peculiarly, the Italian manufacturers, although they have been at the forefront of GP racing since the Continental Circus crystallised in the 1920s, have for the most part shunned production racing. However, a new-found devotion to that wing of the sport took off in the early part of the 1970s and, at the Imola Gold Cup meeting over Easter 1971, a prestigious production race was run as an appetiser to the main event. Benelli entered Phil Read, who was a non-starter, and Charles Mortimer aboard a brace of the new 650 cc twin-cylinder machines. Mortimer was, however, outpaced by the dominant Laverda SFCs of Augusto Brettoni and Roberto Gallina and hence the factory soon pulled the plug on that particular venture.

More successful were the efforts of the British importers, Agrati Sales, who entered their wares in the most demanding of events: the TT. The most noteworthy campaign was that of 1974 when a squad of Murray, Porter and Horton, with 8th, 14th and 16th places respectively, took the 250 cc Production TT team prize aboard virtually stock 230 cc 2C models.

Agrati also entered their four-cylinder 500 cc and six-cylinder 750 cc road bikes, sometimes bored out, in the Formulae races that were introduced to Manxland at the end of the decade, as a sop to compensate for the loss of world championship status for the TT series. The best result was Joey Dunlop's 5th place in the 1978 Formula 2 race.

But, magnificent though these machines undeniably were in many respects, they were motorcycles that raced, not thoroughbred racing motorcycles.

However, the spirit of the family's racing endeavours lived on, transplanted from Pesaro to nearby Urbino, in the guise of Benelli Armi. This new company had been formed in 1967 to make shotguns and, following the de Tomaso takeover, many of the ex-Benelli employees found fresh employment there. The business was guided by Paolo Benelli, assisted by Innocenzo Nardi Dei.

Joey Dunlop astride his six cylinder model in the 1979 Formula One TT *(Sauro Rossi)*

Pat Sproston aboard a 250 cc 2C at Three Sisters, June 1991 *(Alan Horner)*.

Pier Paolo Bianchi on the works 125cc MBA, Silverstone 1980.

POSTSCRIPT

But motorcycles remained at the forefront of Paolo Benelli's ambitions and during the early 1970s no less than 70,000 frames were churned out for use by Benelli, Moto Guzzi and Aermacchi. Paolo hankered after the good old days, bemoaning the fact that he could not produce a motorcycle called Benelli as the name had been sold to de Tomaso, and once claiming: "With us, for de Tomaso to be in control in Pesaro is like having Peron on the throne of England."

The right opportunity came in the form of an alliance with Giancarlo Morbidelli, whose 125 cc twin captured the world title in 1975 and the next two seasons. Morbidelli saw a gap for a production version for sale to privateers and accordingly he came to an agreement with Paolo Benelli that the latter's brand new factory, situated at Sant'Angelo in Vado, a few miles from Urbino, would build the bikes. Hence, Morbidelli-Benelli Armi, or MBA, was born. After a couple of years of success, Morbidelli called a halt to the joint venture, leaving Paolo Benelli to continue the endeavour in the guise of Motocicli Benelli Armi - hence retaining the MBA name.

MBA hit the heights with 125 cc world titles in 1978 (Lazzarini) and 1980 (Bianchi). The machines, which packed the grids, were nothing startling technically, although Paolo Benelli would experiment when necessary, for example collaborating with Ferrari in casting crankcases.

Although a single-cylinder bike was produced by the factory in 1987 for the new 125 cc formula, it proved to be uncompetitive and MBA faded from the scene.

The marque's other challenger was a 250 cc steed, first produced in 1979. Gianpaolo Marchetti and Loris Reggiani campaigned the bike without conspicuous success. The arrival of the peripatetic two-stroke wizard Jorg Muller brought about an improvement, with Roland Freymond winning the Swedish GP in 1982 and Maurizio Vitali taking the Italian title two years later. However, by then the Japanese giants had re-entered the 250 cc arena and so the days of the minnows such as Morbidelli and MBA were over.

The years have, inevitably, taken their toll on the family. Its racing enthusiast, Domenico, passed away in 1975 and the final link with the factory's origins was broken in 1981 when Giovanni died at the grand old age of 91.

The 125 cc MBA formed the mainstay of national and international grids; here with Stefano Caracchi *(Sauro Rossi)*

Works rider Marchetti keeps the 250 cc MBA ahead of Pazzaglia's Morbidelli (*Sauro Rossi*).